WITHDRAWN

THE MIRACULOUS FEVER-TREE

The Cure that Changed the World

Fiammetta Rocco

HarperCollins*Publishers*

For Dan, for the best of gifts and so much else besides
And for my father who, without quinine,
would have died as a boy

HarperCollins*Publishers*
77–85 Fulham Palace Road,
Hammersmith, London W6 8JB

The HarperCollins website address is:
www.harpercollins.co.uk

This paperback edition 2004
1 3 5 7 9 8 6 4 2

First published in Great Britain by
HarperCollins*Publishers* 2003

Copyright © Fiammetta Rocco 2003

Maps and diagrams © Hardlines Ltd

ISBN 0 00 653235 7

Set in Linotype Horley Old Style

Printed and bound in Great Britain by
Clays Limited, St Ives plc

CONTENTS

ILLUSTRATIONS

My French grandmother, Giselle Bunau-Varilla, shortly before she eloped to Africa in 1928. *(Private collection)*

My Italian grandfather, Mario Rocco, in Kenya with my father as a baby in 1930, just weeks before he fell ill with malaria for the first time. *(Private collection)*

Drawing sent to his doctor by Albrecht Dürer, who fell ill on a visit to the malarial province of Zeeland in 1520. *(© Wellcome Library, London)*

The Santo Spirito Hospital in Rome, which in the summer months would fill with patients suffering from the Roman or marsh fever. *(© Wellcome Library, London)*

Malaria, by Ernest Herbert. *(© Scala)*

Three wall-mounted plaques in the former apothecary of the Santo Spirito Hospital tell how quinine was brought to the attention of Jesuit priests in southern Ecuador, how it was introduced to Europe, and how it was distributed to the poor of Rome. *(© Wellcome Library, London)*

The Cambridge doctor Robert Talbor's secret fever cure was found after his death to be largely made up of quinine. *(© The Wellcome Library, London)*

Harvesting cinchona, as it would have looked to Charles-Marie de la Condamine and Joseph de Jussieu when they visited the Peruvian Andes in the late 1730s. *(© Wellcome Library, London)*

Joseph de Jussieu. *(© Bibliothèque Centrale du Musée National d'Histoire Naturelle – 2003)*

José Celestino Mutis, Spanish physician and botanist.
  (© Wellcome Library, London)

Linnaeus, who erroneously named the tree 'cinchona' in his
  system of plant taxonomy. *(© The Linnaeus Museum, Uppsala,
  Sweden)*

Cinchona officinalis, Cinchona calisaya and *Cinchona ovata* as
  drawn by eighteenth- and nineteenth-century travellers.
  (© The Natural History Museum, London)

William Balfour Baikie. *(© Orkney Library Photographic
  Archive)*

My great-grandfather, Philippe Bunau-Varilla, who overcame
  both yellow fever and malaria in the 1880s during the
  unsuccessful French attempt to build the Panama Canal.
  (Private collection)

Sir Clements Markham, President of the Royal Geographical
  Society. *(© National Portrait Gallery, London)*

Markham's exploration of the Peruvian Andes would lead him to
  try to establish cinchona plantations in India's Nilgiri hills.
  (© George Bernard/Science Photo Library)

Richard Spruce. *(© Royal Botanic Gardens, Kew)*

Charles Ledger. *(© Wellcome Library, London)*

Manuel Incra Mamani. *(© Archives of the Pharmaceutical Society,
  London)*

Dr John Sappington. *(© Missouri Department of Natural
  Resources, Arrow Rock State Historic Site)*

A nineteenth-century advertisement for quinine.
  (© Bettmann/Corbis)

Charles-Alphonse Laveran, the first person to identify the
  presence of the malaria parasite in human blood. *(© Wellcome
  Library, London)*

Major Ronald 'Mosquito' Ross, whose discovery that mosquitoes
  were responsible for transmitting malaria would win him the

Early-Eighteenth-Century South America

Central Africa

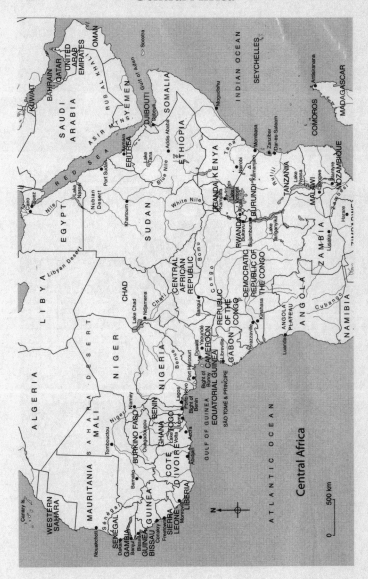

World Distribution of Malaria

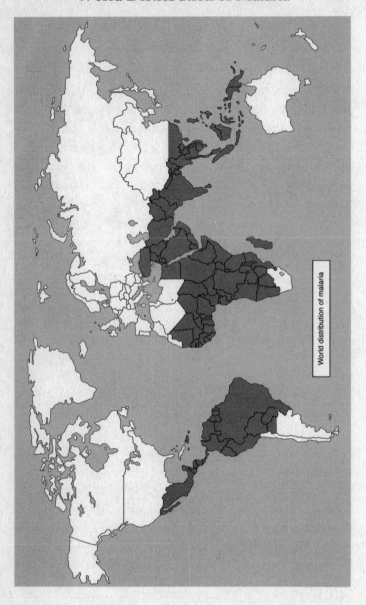

World distribution of malaria

ACKNOWLEDGEMENTS

If *The Miraculous Fever-Tree* had two godparents, they were Roy Porter, Professor of History of Medicine at the Wellcome Institute in London, and Bill Hamilton, Royal Society Professor in the Department of Zoology at Oxford University. Both died before this book was published, Roy while out cycling shortly after his retirement in 2002, and Bill two years before that, of cerebral malaria contracted during a trip to the forests of the Congo, proof of just how dangerous this disease still can be. No one could have been more patient or more encouraging of a non-scientist than these two academics, each a giant in his field and a prince among men. They continue to be deeply missed.

Many chapters of the malaria story have been told before. The challenge of finding something new to say begins with the primary sources – the diaries, letters, notes and inventories that are collected with such care the world over. The librarians of the following institutions deserve a special mention: in London The British Library, The London Library, the Royal Botanic Gardens at Kew, the Royal Pharmaceutical Society of Great Britain, the Wellcome Institute, the Royal Geographical Society and the Oriental & India Office Collections; in Paris the Bibliothèque Nationale de France; in Spain the Real Jardín Botánico in Madrid, Archivo General de Indias in Seville and Instituto de la Historia de la Medicina in Salamanca; in Rome the Biblioteca Vaticana, Biblioteca Lancisiana, Archivio di Stato di Roma La Sapienza, and Archivio della Compagnia di Gesú; in Trevuren, Belgium, the Musée Royale de l'Afrique Centrale; in Panama the Smithsonian Tropical Research Institute in Ancón, Balboa, and the Panama Canal Museum in Panama City; in Lima the Archivo General de la Nacion; in Columbia, Missouri, the J. Otto Lottes Health Sciences Library, and in Nairobi the Kenya National Archives.

Many individuals also offered anecdotes, helped me track down sources or read parts of the manuscript. Jaynie Anderson, Richard Pau and Valeria Rocco in Australia; Simona Bertolli, Father Joseph de Cocq SJ and Dott. Marco Fiorilla in Rome; John Francis Lane in Reggio Calabria; Jane Holligan and Father Juan Julio Witch SJ in Lima, Peru; Dr Saul Jarcho in New York; Etienne Emry, Dirk Gebbers, Horst Gebbers and Elisé Mudwanga in Bukavu, Democratic Republic of Congo; Stanley Munyua, Dr Mauro Saio, Emilio Santasilia and Dr Sally Weekes Seifert in Nairobi, Kenya; Helen Vesperini in Kigali, Rwanda; Jill Watson in Edinburgh; and Rosamund Bartlett, Richard Dowden, Father Francis Edwards SJ, Stephen Fay, Anthony Gottlieb, Steve King, Kyela Leakey, Natasha Loder, Manfred Meysenberg, Eva Monley, Sir Christopher Ondaatje, Mirella Ricciardi, Miranda Seymour, Dr Janice Taverne, Suzanna Taverne and Ann Wroe in London. Some of these are friends, some colleagues; all were unfailing in their generosity.

Particular thanks go to a former colleague at *The Economist*, Edmund Fawcett, from whom I learned so much in the years we worked together, and to Bill Emmott, the editor of *The Economist*, for offering me the nicest job in London. Michael Fishwick made a gesture of great confidence in buying this book on the basis of a five-page letter. I hope I have lived up to his expectations. Robert Lacey's suggestions and advice were those of a peerless editor and a true professional. Brave, generous Toby Eady is a Gurkha of an agent; never had a writer a more faithful friend than he. My last thank you is to my family, especially my husband Dan and my daughter, Tosca, who was just two when this project started and who never stopped asking when it was all going to be stuck together with the picture on the front. Well, here it is.

Fiammetta Rocco,
London, April 2003

Francesco Torti's 'Tree of Fevers'

The Tree of Fevers

*'Cinchona revolutionised the art of medicine as profoundly
as gunpowder had the art of war.'*

BERNARDO RAMAZZINI, physician to the Duke of
Modena, *Opera omnia, medica et physica* (1717)

Francesco Torti's 'Tree of Fevers' may be nearly three hundred years old, but it swells on the page as though it rose from
the ground this very spring. At the crown, its trunk branches
like an earthly anemone, and its arms grow thick and dark.
On the left side of the engraving, the tree bark hums with sap
and leaves grow at intervals in thick bunches of green. The
branches on the right, by contrast, are denuded and leafless.
Tissue-white, they curl upwards as if in supplication to the
Almighty.

Torti, who once saved his own life by taking a dose of
powdered Peruvian quinine bark to cure an intermittent fever,
as malaria was once called, believed that there were two kinds
of fever: those, represented by the leafy branches on the left
side of his tree, that respond to treatment with the bark; and
those, like the dead willowy kind, that do not.

But Torti took his inspiration from another tree, one that he
had never seen, and one that for centuries would remain an
enigma. The magnificent *Cinchona calisaya*, the red-barked
Andean tree that produces quinine, is one of ninety varieties of

cinchona, a relative of the madder family, which also includes coffee and gardenias. Some cinchonas have large leaves, some small; some smooth, some roughly corrugated. But the leaves on the older trees of the true red bark – the *cascarilla roja* that grows eighty feet high – are fiery red. The colour offsets the lilac-like flowers that grow in delicate white clusters, and which are followed by a dry fruit that splits, at the onset of winter, to release narrow winged seeds so tiny and fine that they run to as many as 100,000 to the ounce. Joseph de Jussieu, the first European to set eyes on the cinchona, thirty years after Torti's engraving of 1712, believed that *Cinchona calisaya* was the most beautiful tree he had ever seen.

For Torti and de Jussieu, intermittent fever or malaria was a disease of the Old World. No one knew for certain where it came from or what caused it. But everywhere the Old World expanded its boundaries – pushed ever on by commerce, religion and war – malaria followed. And the price it exacted was beyond imagining. 'Malarial fever,' wrote Sir Ronald Ross, the Englishman who in 1902 was awarded the Nobel Prize for Medicine for proving that malaria is transmitted by mosquitoes, 'is important not only because of the misery it inflicts upon mankind, but also because of the serious opposition it has always given to the march of civilisation . . . No wild deserts, no savage races, no geographical difficulties have proved so inimical to civilisation as this disease.'

Within the foul-tasting, bitter red bark of the cinchona tree is an alkaloid that prevents and treats malaria. The Peruvian bark, which was first brought to Europe in 1631 or thereabouts, was looked upon as a miracle. But its discovery was also a riddle. Cinchona was a tree of the New World. It grew where the rain was plentiful in the foothills of the high Andes, where malaria had never existed. How did anyone guess that among all the trees in South America, it was the bark of the cinchona that would cure malaria? How was it that a seed so small it is almost invisible could grow into a tree, one eighteenth-century source wrote, that was as crucial to the art of medicine as gunpowder had been to the art of war?

This book is the story of the riddle of quinine, the miraculous fever-tree which transformed medicine – and history.

1

Sickness Prevails – Africa

'Malaria treatment. This is comprised in three words: quinine, quinine, quinine.'

SIR WILLIAM OSLER, Regius Professor of Medicine,
Oxford, 1909–17

'If you ever thought that one man was too small to make a difference, try being shut up in a room with a mosquito.'

THE DALAI LAMA, 1977

My grandparents had been married for many years when they left Europe for Africa in 1928, though not to each other.

My Parisian grandmother, Giselle Bunau-Varilla, had had at least two husbands, if not three. My Neapolitan grandfather, Mario Rocco, was being sought by Interpol for trying to kidnap his only child. His first wife, a tall, thin Norwegian with wide cheekbones and a finely arched brow, had been labouring for years to expunge him from her life. She wanted, above all, to change their daughter's identity from Rosetta Rocco, a Catholic, to Susanna Ibsen, a Protestant – and to be rid of her husband forever.

The Neapolitan solution was to remove the child by force and go into hiding, a plan that ultimately failed, though not before it had annoyed the authorities and landed my grandfather in a great deal of trouble.

As an antidote, a year-long safari in the Congo seemed a welcome distraction to all concerned. Yet as the moment of departure drew near, both my grandparents were filled with the excitement of the unknown. Their journey turned from being an all-too welcome respite from their domestic travails to a grand, passionate tropical adventure.

A few hours before New Year 1929, they boarded the sleeper train in Paris that was bound for Marseilles. My grandmother, as always, could be counted on to remain calm even while eloping to Africa with someone else's husband. My grandfather, who had jet-black hair with a deep white streak that swept back from his forehead, only felt his fine sense of the dramatic swell as he put Paris behind him. 'Don't even tell my in-laws what continent I shall be in,' he wrote to his family from the train.

In Marseilles they boarded the SS *Usambara*, a passenger ship of the Deutsch Öst Afrika line that would bear them across the Mediterranean to Port Said, through the Suez Canal, and down the East African coast to Mombasa. From there, the plan was to travel by train and on foot across Africa's thick equatorial waistline to the heart of the continent. They thought they would be away for at least a year. Longer, perhaps.

My grandparents were accompanied by a sizeable quantity of luggage. To equip themselves for a hunting trip that would take them as far west as the Ituri forest on the banks of the Congo river, they had paid a visit to Brussels, to the emporium of Monsieur Gaston Bennet, a specialist colonial outfitter who sold ready-prepared safari kits with everything a traveller might need for a journey of three, six or even nine months.

Monsieur Bennet's inventory sounds much like the

necessities that H. Rider Haggard's hero Alan Quartermain packed when he set off in search of King Solomon's Mines. For their extra-long hunting trip, he sold my grandparents four heavy-calibre rifles, including a double-barrelled Gibbs .500 which my grandfather Mario, with manly Neapolitan excitement, described in his diary as *'una vera arma'* – a real weapon – and a .408 Winchester for my grandmother Giselle, who hoped to shoot an elephant. Eight months later she killed a lone male; its tusks soared high above her head when it lay dead on its side. She allowed herself to be photographed alongside the beast, leaning heavily on the barrel of her rifle as if it were a staff. But the truth is that she felt a little sick at what she had done. Killing the elephant unnerved her. She was five months pregnant at the time, which may have made her especially sensitive. She never shot an animal again.

As well as the rifles, my grandparents were outfitted with two pairs of shotguns, a twelve-bore and a lady's twenty-bore; five hundred kilos of ammunition in watertight boxes; six trunks of tropical clothing; twelve cases of brandy; eight of books; a typewriter; a gramophone with my grandfather's favourite record, 'My Cutie's Due at Two-to-Two Today'; coloured beads for gifts; and enough sketchpads, pastels and modelling clay to last them a whole year – my grandfather was a painter and my grandmother a sculptress. Their effects were packed into tin trunks weighing not more than twenty-five kilos each, the maximum that would be carried by an African porter. Giselle stood barely an inch over five feet and always wore a turban, which had the effect of both hiding her in-cipient baldness and making her seem taller than she really was. When my grandparents reached the Ituri forest she unpacked her clay and set about modelling a local Tutsi chief who towered nearly two feet above her. He watched her as she worked, his face impassive. He said nothing, but his children danced around and called her 'Potipot', she who works with clay.

In addition to the safety precautions of heavy Damascus-barrelled guns and several changes of boots, Monsieur Bennet packed my grandparents a sizeable medicine chest that was manufactured from black metal and lined with marbled endpapers to absorb any moisture and keep its contents safe from ants. In it he placed gauze bandages and sutures, several bottles of Dr Collis Brown's Elixir, a concoction made of morphine, cannabis and treacle that had been invented in 1856 and was recommended for treating diarrhoea, boric acid for the eyes, carbolic acid against lion and leopard scratches, Epsom salts and castor oil for constipation, and a brown goo called Castellani's Paint to fight skin fungi. There were also twenty-four sets of steel syringes and needles, each packed in a small metal box with a tight lid for easy boiling, the best method of sterilisation in the bush. No medicine chest bound for Africa was complete without a supply of purple crystals of permanganate of potash, for washing raw vegetables and cleaning out snakebite wounds. With it came a snakebite pencil which you used to cut a Y-shaped incision, so you could lift the skin immediately surrounding the bite and pack it with permanganate.

Snakes are highly sensitive to vibration, and most of them will slither away when they detect you approaching. Mosquitoes, on the other hand, do not. Among the most important items in Monsieur Bennet's medicine chest was packet after packet of powdered sulphate of quinine, to guard against malaria. Alan Quartermain packed an ounce of quinine and one or two small surgical instruments into his bag for the final assault on King Solomon's Mines. He would not have left home without it. 'This was our total equipment, a small one indeed for such a venture,' he wrote. 'Try as we would we could not see our way to reducing it. There was nothing but what was absolutely necessary.'

From Mombasa Mario and Giselle headed west towards their first stop, Voi, a railway junction halfway between Mombasa and Nairobi. The land was flat and scrubby, with occasionally a mound of hills rising in a greeny-purple haze in the far distance. They saw Masai herders with thin, high-boned cattle that were oblivious to the sun's heat. Dried-out umbrella thorns provided the only shade, and a patchy shade at that. Shortly after Voi they made a detour south across the border with Tanganyika to try to get a better view of Mount Kilimanjaro, Africa's highest peak. They passed the spot where General von Lettow-Vorbeck, the German soldier-adventurer, had routed a British regiment fourteen years previously. By 1917 the British had begun to fight back, and von Lettow was in trouble. Supplies were running short. Recurring bouts of malaria had reduced many of the soldiers, von Lettow among them, to yellow, shrunken skeletons. Unable to obtain any imported quinine tablets, von Lettow's officers began making it themselves from the powdered bark of cinchona trees that they found growing locally. The cinchona had been planted in the early 1900s, by Tanganyika's German colonial masters. Von Lettow's soldiers couldn't make tablets, though, so they stirred the ground-up bark into their coffee. It was a horrible brew which the troops called 'Lettow-schnaps', but it worked.

Although they had lived in Europe their whole lives, both my grandparents already had some experience of malaria before they left for Africa. In 1886 my great-grandfather, Philippe Bunau-Varilla, became the Chief Engineer of the

Panama Canal, a scheme that had been dreamed up by Ferdinand de Lesseps shortly after he had finished his canal at Suez. By the time France's Panama project collapsed in 1889, twenty-two thousand men had died of yellow fever and malaria.

No one made it a requirement that those who went to Panama should take regular doses of quinine. This is astonishing, for quinine was already well known by then – Jules Verne wrote about it in his novel *L'Île mystérieuse* in 1874; later Chekhov would call his favourite dog Quinine (being a doctor, he called his other dog Bromide). The problem was that quinine was difficult to obtain, as supplies from the 1860s on were intermittent. Worse still for the project's managers, it was expensive: while the American Civil War was at its height, much of what was available was shipped north to protect the Union soldiers who were taking over more and more of the Confederate land where malaria had long been a scourge, and that trade still ran strong after the war ended. The officials of the French Compagnie Universelle du Canal Interocéanique calculated that it was cheaper to let its workers die than to spend a lot of money trying to cure them with costly medicines. Even a prophylactic dose, which would surely have saved them much money over the long run, was, they calculated, beyond their budget. The Americans, who took over the canal's building works in 1903, were of the opposite view, and forced their workers to take a regular prophylactic dose of quinine or face mandatory punishment. In less than a year, the US Army's soldier-engineers managed to stamp out virtually every trace of malaria. But that is getting ahead of the story.

My grandfather, for his part, was born in Naples but spent much of his childhood staying with an aunt who lived in the hills of Maremma. To many, this part of Tuscany was known as '*la Maremmamara*' – bitter Maremma – because of how malaria had forced people to abandon the land in the eighteenth and nineteenth centuries. Another aunt lived in the Roman Campagna, where malaria had existed since Roman times, and from which it was not wholly stamped out until the 1930s, when Mussolini embarked on draining the Pontine marshes at the mouth of the Tiber, thus ridding western Italy of the pools of stagnant water in which the malaria-carrying mosquitoes bred.

In truth, the whole of southern Italy in summer was a hellhole of malaria. Travelling through the region in 1847 on his way to Sicily, Edward Lear, the artist and poet whose children's verse usually speaks lovingly of the oddities across the seas, noted in an unusually serious vein that malaria turned the population yellow and shrivelled many to living skeletons. 'After May,' he wrote in a letter to his brother in the spring of that year, 'the whole of this wide and fertile tract . . . is not habitable, and in July and August to sleep [i.e. to die] there is almost certainly the consequence of fever.'

George Gissing, who made the same journey nearly sixty years later, wrote in his Calabrian classic *By the Ionian Sea* of the amiable Dr Sculco, who advised him to 'get to bed and take my quinine in *dosi forti*. [Was I not] aware that the country is in great part pestilential [because of] *la febbre?*' Of course, Gissing, Lear and the other foreign writers who journeyed to the south of Italy could always leave if things got too bad. For the innkeeper in Giovanni Verga's nineteenth-century short story 'Malaria' there was no such option. First came the railway, which took away the brisk business he'd enjoyed from the carriage trade. Then it was the malaria that struck, bearing away each of his four wives in turn, earning him the nickname 'Wifekiller'. When none of the village girls would consent to become his fifth bride, he said to himself, 'Next time I'll be taking a wife who's immune to the malaria. I won't go through all this again.' But it was not to be.

'The fact is,' wrote Verga, 'that malaria enters your bones with the bread that you eat and whenever you open your mouth to speak . . . The malaria fells the townspeople in the deserted streets, it pins them down in the doorway of houses whose plaster is peeling in the sun, as they shudder from the fever, wrapped up in their overcoats, and with all the blankets from their beds round their shoulders.'

Massimo Taparelli, the writer and statesman who, as Marchese d'Azeglio, served as Prime Minister of Italy under King Victor-Emmanuel II, often mentioned the disease in his diaries. 'While we were staying at Castel Gandolfo [the Pope's summer home],' he wrote on one occasion in 1860, 'I used to go down to the plain to shoot. But instead of birds I got the terrible marsh fever, the ancient scourge of Latium . . .

'No one can have any idea of the iciness of the cold phase or the burning heat of the hot attack of these painful fevers. Quinine is certainly the most beneficent discovery for the Roman Campagna. There may be no steam there, no newspapers, no other modern inventions but at least they have quinine, and that's worth all the rest put together.'

When my grandfather was growing up, everyone in southern Italy regularly took quinine in the summer, when the danger of catching malaria was at its worst.

My grandfather saw his African safari as just the start of a grand adventure that would take him and my grandmother around the world. 'From here, we shall travel on, to Dar es Salaam, to Beira and then around the Cape to Rio de Janeiro,' he wrote to his mother as they arrived in Africa in February

1929. They never left. Later that year my grandmother, who had already lost her first baby in childbirth, became pregnant again. Already thirty-seven, she wanted to take no risks a second time. The couple returned to Nairobi to await her confinement.

While she was in hospital, Mario took off in a small plane to look for a friend they had made during their months in the Congo. Before the end of the day he ran out of fuel and crash-landed by the shores of Lake Naivasha, about seventy-five miles from Nairobi on the shady floor of the Rift Valley. A grizzled Englishman, who introduced himself as Harvey, hailed him when he climbed, unhurt, from the wreckage. Mr Harvey took him back to his house, a bungalow with a mottled thatch roof, where after several stiff drinks and a lot of talk he offered to sell Mario his property. Hurrying back to Nairobi to tell Giselle, Mario stopped at the telegraph office to send a cable to his father-in-law, who would be putting up the purchase price.

That telegram was sent more than seventy years ago, and I have it here before me as I write. Its blue folds are as soft as a baby's cheek, and the pages quite floppy with being taken out and put away so many times. It is addressed, in brief telegraph-speak, to 'Bunovarila, 1 Grande Chaumiere, Paris'. And it says: 'Purchased shamba Naivasha 3000 acress [sic] three miles lake front. 5000 pounds. 2000 cash, balance three years. Best bargain. Cable if you want me home to fix everything or cable approval.'

By the time Mario and Giselle decided to stay in Africa on the farm by the shores of Lake Naivasha, the supply of little packets of quinine sulphate they had brought with them in 1929 had long since been used up. Neither of them had caught malaria in the Congo; only heatstroke. But in 1936

Mario came down with a bad attack as he returned from a trip to Lake Victoria, a notorious malarial spot even today. In 1940, just after the start of the Second World War, he had another attack while he was in a British internment camp in Nairobi.

As an Italian, Mario had been arrested as soon as war broke out in September 1939. Giselle, a French national, was allowed to remain on the farm, where she turned her attention from sculpture to raising pigs, as well as to my father and his two sisters. She stayed in touch with her family in Paris, and though the seaborne post was slow, it did its work. Once a month, sometimes more often, a postal vessel docked at Mombasa and a few days later my grandmother received a delivery of letters, newspapers and parcels from Europe, which contained among other things regular supplies of quinine for all the family. And, if she was lucky, there might also be a letter from my grandfather.

I found those letters in an old shoebox the morning after Mario died in 1976. For more than a quarter of a century he had kept them under his bed. Lonely, frustrated and often sick in the internment camp, he longed for home. His letters were restricted by the camp authorities to a single sheet of paper, and over time he perfected the tiniest, neatest handwriting you ever saw, so that he could write first in one direction and then at right angles over the page, stretching out for as long as he could the connection with his family. 'I dreamed last night that you were sitting by my bed,' he wrote to Giselle in the winter of 1943 after a bad attack of malaria. By then he had been interned for nearly four years. 'Nothing would heal me more quickly than to feel your hand upon my cheek.'

Through the war years Mario fell ill several times with

malaria. 'I don't know what is worse; the fever or the shivering,' he wrote. 'There is no quinine. Cold water is the best we
can hope for.' Beyond the heartache and the loneliness there
was a cold reality about malaria in Africa that is as relevant
now as it was then. With access to efficient anti-malarial
drugs, Giselle and her children remained healthy. Mario, who
like so many Africans today did not have the medicines he
needed, did not.

Back in Kenya in the early 1950s he had a third bad attack,
and in 1958 a fourth while on a long winter visit to Europe.
After that, it would often strike when the rainy season was
under way. Our farm, with its warm climate and its clumps of
thick papyrus that stretched out for yards into Lake Naivasha,
was the perfect habitat for the *Anopheles* mosquito that
spreads the disease. In the rainy season, when the mosquito
larvae hatch in their thousands, it can be especially bad, and
even today we always sleep under mosquito nets.

I have had malaria only once, when I was eighteen. I had
been on holiday at the mosquito-ridden Kenyan coast, and
cared little about remembering my pills. That was enough.
Soon after I returned, I began feeling unwell. I took my
temperature. 101°F. By nightfall it was up to 104° and I was
beginning to hallucinate. With any other illness, I have
always felt that I was still myself. I might be in pain or feel
nauseous, but I was me – only sicker. Sick with malaria,
however, my body felt it was no longer my own. It had been
invaded, as if it had been subjected to a military coup. I
remember walking into my father's bedroom; I watched
myself, as if I were another person completely. The fever was
just beginning to shoot up. The parasites in my blood that had
invaded the red corpuscles were splitting them open and
destroying them in a rampant urge to reproduce. I lay down

11

on the bed, and passed out. After that everything is blank. My blood had been hijacked. That is how the delirium begins. 'I have lain on my cot for forty days,' the explorer David Livingstone wrote to his wife from Luanda, in present-day Angola, in 1854. 'So fierce was the delirium that I remember almost nothing of it.' The fever would kill him nearly twenty years later. Clearly, I was lucky.

My father gets it more often than any of us, and worse. Just a few days before I wrote this, he called to say he was ill again. 'I began to feel colder and colder and colder,' he told me, his voice thin with fever. 'I got into bed with a hot water bottle and kept piling on blankets. For two or three hours I just shivered and shuddered as if I was in an icy blast. Then, suddenly, it stopped. And I started getting hotter and hotter and hotter, and throwing all my covers off. Forty-eight hours later it started all over again. And every forty-eight hours it's been the same for about a week.'

As always, my father went to his Italian doctor in Nairobi, Mauro Saio, one of the world's leading specialists in treating malaria. Dr Saio has worked so long with the disease that he named his speedboat *Anopheles* after the mosquito that spreads the disease. 'You have headache, vomiting, diarrhoea,' he explained, 'and if it's not caught in time and the parasites keep reproducing, you can have respiratory distress and systemic organ failure.'

For Dr Saio, combatting malaria is a campaign. As he told me the first time we met, 'It's a battle. A hard battle. I know this disease. I fight this disease every day of my life. It is my personal enemy.'

My grandparents tried to protect themselves and us, my sister and my four cousins. As far back as I can remember, the daily ritual of breakfast on the farm was broken on Sundays by the distribution of the quinine, or its modern chloroquine-based equivalent, Nivaquine: two tablets for the grown-ups, and for the children a spoonful of Nivaquine syrup, which was increased to two spoonfuls when we were about twelve years old. Oh, it tasted awful. It wasn't like today, when pharmaceutical companies try to make their medicines palatable to children; in the 1960s they had other priorities – all a medicine was required to do was to work, and you just had to take it. Quinine is marked by its particularly bitter taste. Over the centuries, many people have refused to swallow it for fear that they were being poisoned.

Nivaquine is also bitter, and the vile taste of the syrup clings to your teeth and gums long after you have swallowed it down. Just writing about it makes me wince at the memory. It tasted so ghastly that my grandfather had to devise his own method for persuading us children to take it. He bribed us. If we swallowed down the Nivaquine, we were allowed to choose what we would have for Sunday lunch.

This was no mean bribe, for my grandfather was a tremendous cook. By his place at the head of the table lay a book covered in well-loved, shiny dark red leather. *Il Talismano della Felicità* was written nearly a hundred years ago, and it contains instructions for making every manner of Neapolitan delicacy. Once we had all swallowed our Nivaquine, my grandfather would pour himself another cup of black coffee, drop into it a lump of sugar, light a cigarette and then reach for his *Talismano*. Slowly he would turn the pages, stretching out the agony of anticipation. And then he would begin, in a deep, sonorous voice. 'So, *bambine*, what will it be today? *Pizze fritte? Sartù di riso? Maccheroni al ragù? Melanzane alla parmigiana?*' We would vie to be the one who made the final choice, completely forgetting the filthy taste of the Nivaquine in our anticipation of the meal to come. In our house, the danger of malaria was vanquished by greed.

When I was fourteen, I was sent to boarding school in England. I arrived at my new convent school in Sussex on a bleak January afternoon. Snow-filled clouds hung over the landscape like a laundry bag waiting to burst. One of the Irish nuns showed me into a dormitory with seven beds covered with old rose-coloured candlewick bedspreads. Her manner was brisk, and she didn't stay long. I had arrived in the middle of the day in the middle of the school year, and she had things to be getting on with. The other girls were in class, and every bed in the dormitory had been taken except one, that stood alone in the middle of the floor. Slowly I unpacked my trunk, and stowed away the clothes my aunt had ordered off a long list from a department store in central London: thick white underpants (inner, changed daily), huge navy-blue serge underpants (outer, changed weekly). I thought of running barefoot in the soft African dust and splashing in the ditches by the side of the farm roads, and felt a bit sick with the longing to be back home. None of my room-mates, it turned out, had ever been to Africa. They giggled among themselves and argued endlessly about the merits of rival pop stars. At night, they tossed and mumbled and farted in their sleep. There was not a moment of privacy. We even had to share baths. By the end of term the seven of us had been living so closely together for so long that our menstrual periods all began and ended on the same day. But that did not bring us closer. The loneliness of living in a foreign crowd so far from home was with me always. I felt that I had landed on another planet. There was something about the fact that my family, my entire tribe, had packed up everything it owned and turned its back on Europe that set me apart.

14

It was as if I had lived my entire life in another language.

As the winter wore on through February and the windy weeks of March, I felt as if it would never end. I missed my sisters and the mental shorthand we assumed together because we had always lived in the same house. I missed the tropical rituals: barbecues at Christmas, snow that came in tins for spraying on the Christmas tree, and the way the sun went down every day at the same hour, whatever the season. I even missed the beastly Nivaquine, for the danger that forced us to take it was something familiar to me. I missed my grandfather's sweet tomato sauce, and the smell of the land after it had rained. I missed everything so much that I would lie awake at night trying to conjure up the smells of home. It was as hard as sewing raindrops.

Then one day, on one of the rare weekends we were allowed out, a friend of my father's took me on a long Tube journey to St Joseph's Foreign Missionary Society in Mill Hill, in the very outer suburbs of north London, where he had to pick up a package. St Joseph's was the male equivalent of the convent school I attended. But while my Irish nuns had devised a whole book of rules for keeping us from talking to boys or fraternising in any way with the outside world, St Joseph's positively encouraged a spirit of independence in its young men.

During the time that my father's friend concluded his business, I wandered down a long corridor, the walls of which were covered in small photographs of all the priests who had ever served St Joseph's abroad. The early ones were sepia-tinted. Gradually they became black-and-white. It was there that I realised for the first time that pulling up your roots and embarking on a new beginning was something people had always done. Though they knew they might never return, like my grandparents those priests had rolled the dice and boarded ship.

St Joseph's Foreign Missionary Society, or the Mill Hill Fathers as they are commonly known, now has missions in nineteen different countries, stretching from Brazil to St Helena and from Brunei to Sudan, which is impressive in our secular age.

At the end of the corridor was a large wooden door which led to the Society's chapel. On a stand to the left of the altar I found a red leather-bound Roman missal. It was open at the page of prayers for persecuted Christians: 'Father, in your mysterious providence, your Church must share in the sufferings of Christ your son, give the spirit of patience and love to those who are remembered for their faith in You, that they may always be true and faithful witnesses to your promise of eternal life.'

Faith was what had inspired the Mill Hill Fathers to go across the seas. Yet when I sat in the pew of their little chapel and looked around the walls, I realised that it was not a weakening of belief, nor persecution, that killed their priests so often and so young – but disease: overwhelmingly malaria. Running along the top of the wall, just under the eaves, was a stone course carved with the words: 'Pray for the Souls of our Dear Brethren, the Diseased Missionaries of St Joseph's Society.' Beneath it, covering all four walls, were stone slabs listing the names of the hundreds of missionaries who had died in the course of doing God's work. The first slab covered 1872 to 1905, the years of the Society's earliest ventures abroad. On it were the names of thirty-two men, starting with the Reverend Cornelius Dowling, a doctor as well as a priest, who died of malaria in Baltimore on 9 August 1872, at the age of thirty-one.

Like the Reverend Dowling, two-thirds of those commemorated in the chapel had died before they reached the age of forty. They succumbed, far from home, in India, Kashmir, Borneo, Italy, France, Uganda and Singapore. As the years progressed, the ever-expanding array of places where the Mill Hill Fathers passed away is proof that sickness and death did nothing to quench their Christian fire. By 1917 the missionaries were dying in the Congo, Borneo, Uganda, the Philippines, the Punjab and on the Isle of Wight. By 1925 it was Sarawak, Madras, Kisumu in western Kenya, and in the dry north of the country, the bleak and lonely Kavirondo Gulf. The dangers they faced were multiple, yet according to the records, fully three-quarters of them died of the same thing: Roman or intermittent fever, tertian ague, or as it later came to be known, malaria.

It took a special kind of courage to leave home and travel to Africa in the nineteenth and early twentieth centuries. Not only were the distances enormous and the prospects of return uncertain, but Europeans thought of Africa as a kind of wild and unpredictable beast that had to be beaten into submission physically, morally and politically.

There were dangerous animals, savage tribesmen and, always, the threat of disease: sleeping sickness, river blindness, yaws, leprosy, trachoma, typhoid, tick fever, filariasis, beriberi, bilharzia, kwashiorkor, rinderpest and East Coast Fever were just a few of the ailments waiting in Africa. Of the many illnesses threatening both man and beast, though, none seems to have preyed on travellers' minds as much as that which became known in many parts as the 'pioneer shakes' – malaria.

Some diseases were terrifying simply because they were

17

deadly. Yellow fever and malaria's cousin, blackwater fever, which turns your urine the colour of dark Burgundy and your kidneys into fragile sacs that can burst at the slightest movement, are like poisonous snakes: they kill in a matter of hours. But there is something particularly insidious about the way malaria stalks its victims, the way its parasites lurk within the body, hiding from its immune system and lying silent for years until you think you have finally shaken it off, only to find that it always returns, driving you mad with fever, shivering, delirium and pain, weakening you more with every bout before, often, it eventually kills you. As the malaria parasite reproduces in your blood, it swells and bursts out of your red blood cells, leaving in its wake a sludge of wrecked haemoglobin. Some of this material ends up in the liver and the spleen, causing them to swell and turn black. In the unlucky few the parasite accumulates in the capillaries of the brain, causing the cerebral malaria that kills so fast.

Of the thirty-six girls in my primary school class in Kenya, eleven were dead before the age of forty. Five were killed in car accidents, most of them by hit-and-run taxi drivers, who are paid by the journey and drive as fast as they can. One died in childbirth. But four died of cerebral malaria, caused by the deadly *Plasmodium falciparum* parasite, which kills so many people in Africa. Perhaps, as I had done when I was eighteen, they had become cavalier about the dangers, and didn't take their anti-malaria tablets; perhaps they were just unlucky, and failed to get adequate medical treatment in time. Between them they left nine orphaned children.

Daily life in Africa is so harsh that there is often little time to dwell on the nuances and inequities of history. Uppermost in the minds of Europeans who travelled there, from the earliest years, was how to overcome disease before it overcame you. Many who live there today still think of malaria as a ghostly presence in their lives; something that visits and revisits with the advent of the rainy season, and from which you never quite escape. In the nineteenth century Henry Morton Stanley, who reckoned he caught malaria more than two hundred times during his exploring years, carried his own cure, which he called a 'Zambesi Rouser', made of powdered jalop, calomel, crushed rhubarb and quinine, 'to be taken with a little water whenever an attack of malaria threatens'. My father is more circumspect. He takes his weekly pills in silence, and only ever talks of having a 'touch' of malaria, or even a 'go' of it, as if loudly to invoke a more severe diagnosis might in some way be calling down the fury of the fates.

Fourteen miles from my grandparents' farm, on the other side of Lake Naivasha, is the small district hospital. Thirty beds are divided between three wards, but in the rainy season, when malaria can reach epidemic proportions, patients have to queue up to be admitted. Even in the dry months a steady stream of people, most of them women with small children, line up at what passes for an outpatients department round the side of the hospital. Most of them will have travelled in a hot bus or walked many miles to get there, and they sit, uncomplaining and undemanding, beneath the sprawling pepper trees while they wait, sometimes for hours, to be seen by a doctor.

'No wonder they're called patients,' laughs a nurse holding a blood sample. She appears cheerier than she ought to be, considering the long hours of work that still lie before her. She and three doctors will see about 180 patients in a morning, spending enough time with each to give a quick diagnosis, offer a prescription or decide if further examination is needed. There is none of the smart whiteness of Dr Saio's office at the main hospital in Nairobi, though the work that is done here is very similar.

The hospital in Naivasha is run by local community doctors. The consultation rooms are spotless and the walls are papered with educational posters about AIDS, safe sex and the importance of using clean water for mixing infant formula. Everyone pays fifty US cents to see the doctor, the same again for a blood test, and between ten cents and a dollar for medicine. A limited range of drugs is supplied cheaply by the Anglican Church which, despite its charity, is Protestant enough to have concluded early on that people value something more if they have to pay for it, no matter how small the sum.

Outside the door is a hand-painted sign with a message from the first Book of Peter, a reminder that so much in Africa is still a matter of faith. 'Cast all your cares unto Him, for He cares for you,' it says. Cheap and simple to run, the clinic is more effective than one might think, given its simple furnishings and tiny annual budget. For many Africans, this is the very best medical knowledge they will encounter.

A woman in a red patterned skirt and a white headscarf enters the consulting room. Asked what her name is, she mumbles 'Grace' in a barely audible voice. She complains of a swollen stomach. A nurse palpates her abdomen, and concludes that she is about twenty weeks pregnant. Although this would be her third child, Grace seems not to have noticed that her menstrual periods had stopped, or had any idea that she might be expecting. Perhaps another child was too much of a burden for a poor family, and she did not want to admit the truth. The nurse signs her up for admission to the hospital five months hence, and arranges, meanwhile, for fortnightly antenatal visits.

The next patient, Joseph, complains of chest pains. He has chronic oedema. His lower legs look like tree trunks and he suffers from high blood pressure. He pulls his thick jacket around him as the doctor prescribes a new medication for his angina, and shuffles out.

A heavyset young woman in red flipflops and a blue head-scarf comes in next. She speaks softly to the doctor in Kikuyu. She is called Sandra. Both her children are running a tempera-ture and she has a bad chesty cough. She wants them all to be tested for malaria. The doctor examines them. 'Say "ah",' he commands, peering down the throat of each child.

The thick white ulcers of oral *Candida* indicate that they are probably both HIV positive. Without proper medication, it will only be a matter of time before they have AIDS. On the wall is a poster of a strip cartoon showing how AIDS is transmitted. It says nothing about foetal transfer of the virus. Beside it another chart outlines how to prescribe Amodiaquin, the standard treatment for malaria now that chloroquine, a synthetic anti-malarial compound developed during the Second World War, is so ineffective that many African countries, including Kenya, have discarded it. For a baby of less than seven kilos, you give a quarter of the daily dose. For a child weighing more than fifty kilos, the daily dose is three tablets.

The nurse asks each child to put out a hand. Gently she swabs a finger, pricks it and smears the gentle swell of blood onto a glass slide. Moments later a lab technician dips the slides into staining fluid, dabs the end with a piece of kitchen towel to clear the excess moisture, and puts the slide to dry on the warm back of a paraffin picnic fridge beside his desk. In a few moments the slides are ready and he slips them under the microscope, the only piece of machinery in the clinic that runs on electricity.

The circular-shaped parasite, with its dot-like red eye at one edge that is so characteristic of malaria, is clearly visible. Sandra and her two children all have malaria, though they are lucky they do not harbour the deadly *falciparum* parasite. A pharmacist counts out a tiny handful of white pills and slips them into a small square envelope. They are quinine sulphate, which is made from the bark of the cinchona tree grown in the last cinchona forest, in the eastern Congo.

My grandparents may have been unusually adventurous in the way they happily traded in a comfortable life in Paris for an unknown future in Africa, but their caution in insisting that we all regularly dosed ourselves with quinine was proved right. To many Western travellers today, malaria is something that exists over the horizon. It does not carry the slow promise of death that is embedded in AIDS; in this part of Africa, AIDS has seeped into so many villages that small children and old grandparents are often the only people still to inhabit the silent thatched huts. Nor does malaria conjure up an explosive, primitive fear, like being attacked by a lion or bitten by a poisonous snake. Most travellers know that malaria exists, but they buy an ordinary over-the-counter dose of prophylactics and go on holiday regardless, often ignorant of whether the prophylactics work or not. In Britain there are more articles in medical journals devoted to the depressive side effects of mefloquine, or Larium as it is usually known, one of the strongest anti-malarial prophylactic drugs on the market, than on the disease itself.

Malaria stalks Africa, where it is a real cause of fear and grief. The United Nations World Health Organisation estimates that as many as five hundred million people are infected by the disease every year. That is eight times the population of France or Great Britain, or twice as many people as live in the United States.

Of those who fall sick, as many as three million die every year. The very large majority of these are small children for whom clean water, decent food, antibiotics and quinine-based drugs to fight the onset of the disease, let alone a decent pro-phylactic, are no more than a dream, perhaps heard of, but unattainable. Malaria is so common, and so deadly, that the WHO estimates one person dies of it every fifteen seconds. In the last decade it has killed at least ten times as many children as have died in all the wars that have been fought over the same period. Yet the mosquito that carries it is little larger than an eyelash.

Out of just under five hundred different varieties of *Anopheles* mosquito that are recognised today, only about twenty are thought to be seriously responsible for spreading the disease to humans. The malaria parasite packs the salivary gland of the female mosquito, of no danger to anyone includ-ing its host until it bites a human being. Only when it injects some of its saliva containing the malaria parasite into the bloodstream does the mosquito transfer this dread disease. In the course of the bite it also withdraws blood. Its victim may already be infected with the parasite. If the mosquito moves on to other people and bites them, the endless cycle of infec-tion and reinfection will simply repeat itself. The *Anopheles* mosquito needs blood to lay its eggs, but the damage it inflicts on humans is completely incidental to the insect. 'A man thinks he's quite something,' the American writer and car-toonist Don Marquis had his cockroach hero Archy say in *archy and mehitabel*. 'But to a mosquito a man is only a meal.'

The mosquito breeds in pools of stagnant water – overflows from rivers that have flash-flooded and then subsided, roadside ditches, forgotten furrows in uncultivated fields, water butts in towns and rain-filled puddles in the middle of country roads. In the Naples of my grandfather's youth, the mosquito found a comfortable home in the well-watered window boxes of the city tenement buildings. When my great-grandfather was in Panama, it was customary for the nurses in the little French clinic on the hill above the engineering works to stand the hospital beds in huge flat bowls of water to stop the black spiders from climbing up the bed legs and biting the patients. No one could have devised a better breeding ground for mosquitoes had they tried. Among the canal workers of the mid-nineteenth century it was customary to warn newcomers that if you didn't have malaria when you went into hospital, you would undoubtedly catch it while you were there.

Today malaria is chiefly a danger to people in the tropics, particularly the poor, who live in bad housing with inadequate drainage and no mosquito nets, insecticide sprays or fancy prophylactics. But once upon a time it was common all over Europe. Even so, no one knew exactly what it was. Nor did they know how to treat it. When a cure finally was discovered, it revolutionised theories of medicine and the way physicians thought about treating illness.

Nowhere in Europe was the scourge more deadly, and the need for a cure more acute, than around the Basilica of St Peter's in the centre of Rome, where every summer for centuries it killed hundreds of people, making no distinction between peasant, priest or pope.

2

The Tree Required – Rome

'When unable to defend herself by the sword,
Rome could defend herself by means of the fever.'

GODFREY OF VITERBO, poet, 1167

Giacinto Gigli lived for sixty-five years in an alleyway by the Via delle Botteghe Oscure. His small townhouse was one of several built close together. They clung, like a gaggle of shy children, to the end of one of the crooked passageways that cluttered the centre of Rome, seeming almost to lean into one another when the wind whipped around them during the early days of winter. Gigli was not born there, but from the time when he was twelve years old until just before Christmas 1671, when he died at the age of seventy-seven, he returned every night from his work in the papal palace at the Vatican and climbed the stairs to the study, his favourite room, at the top of the house, where his desk was placed between two tall windows.

One of them looked westwards, over the bend in the Tiber where tradesmen, prelates and visitors would cross the bridge that led towards the Santo Spirito hospital and beyond it to St Peter's and the palace of the Vatican. From that window Gigli could just see the golden dome of St Peter's, though often it seemed almost to fade away in the summer haze, when the city became too hot to bear and the Pope's court moved to the palaces and villas in the coolness of the hills that surround the city. The other window, on the far side of the room, looked north-east towards the Quirinale, the highest of Rome's seven hills, where the Pope's own summer palace had been built so that he could escape the unhealthy summer air, and where a gentle breeze blew all day through the shady trees.

The Gigli family was part of Rome's *petite bourgeoisie*, though by his death Giacinto's father had been able to leave his son the property near the Via delle Botteghe Oscure, two other smaller houses in the centre of the city, and a vineyard on the road to Frascati, a considerable inheritance for a modest man. When he was twelve Gigli entered the Collegio Romano, the Jesuit school where he studied grammar, humanities and rhetoric. For a while he studied law in the studio of Angelo Luciano, a well-known Roman advocate, and graduated as a specialist in papal law. By that time his father was dying, and as Giacinto prepared to take his place as the head of the Gigli family, the young man's thoughts turned to marriage. He had known Virginia Lucci all his life. She was his neighbour's daughter, and the time had come to ask for her hand.

As he grew older, Gigli began to acquire the status of a man of respect. He was made a *rione*, representing his parish of Pigna in the committees and on ceremonial occasions that would occasionally bring together the city and the Holy City. Dressed in velvet and bedecked with feathers, he would walk in line before the horsemen that accompanied the papal processions. Later he was made a *caporione*, before eventually serving twice as the *priore*, the head of all the *caporioni*, responsible to the Pope for helping collect taxes and keeping order within the Holy City. These duties gave Gigli an intimate insight into Rome, and particularly the clerical administration that ran it. He became something of an expert on Vatican politics – who was in favour and more importantly who was not, how different cardinals behaved and how they were discussed. 'All this and more is the lifeblood of the Holy City,' he would write.

We don't know what Gigli looked like, for no contemporary likeness of him survives. But we know a great deal else about him: how he lived, how he filled his day, what he did when he fell ill, whom he saw and what he ate. Gigli was meticulous about recording the details of his life in his diary, which survives in the Vatican library, writing something every day, even if only a brief phrase or two. He wrote in Latin, in an elegant slanting hand, taking care to reach the very edge of the page before starting on the next line.

Gigli had acquired the habit of chronicling his daily life before his marriage. In 1614, when he was only nineteen, he even began an autobiography, entitled *Vita*. He kept this up assiduously for about five years, making minute notes of everything that went on in his household, recording the names of the servants who came and went, what they were paid and what they earned in tips or small gifts, like the woollen socks that were given to the nurse who came to care for his only son. While this might seem unnecessarily fussy to some, these were not bad habits for a diarist to acquire. In addition, he also wrote poetry, long verses about his native city and eulogies of the Popes in rhyming octets.

The papal court was the centre of Gigli's life. From the first entry in the diary, on 29 May 1608, to the last, when he was almost too blind to write any more, yet fretted about missing the baptism of the Pope's new baby niece, he specialised in the comings and goings of the Vatican. The Pope's court was newly returned to Rome after the alternative papacy, set up the previous century in Avignon with the support of the King of France, had threatened to rob Rome of much of its wealth and influence. Restored once more to its traditional home, the leadership of the Roman Catholic Church was doing all it could to spread the counter-Reformation in Europe and impose its spirit, burning heretics and attempting to suppress Protestantism whenever it could.

In the first decades of the seventeenth century, though, when Gigli was making his daily observations of the Holy City, papal Rome still hadn't really found its feet. The memory of the schism lived on, making the city fathers nervous, hidebound, inward-looking; too fearful of the Turkish forces that threatened Venice, or the Spaniards who lay siege to Naples, ever to be easy. Naturally conservative, papal Rome then was more fearful of change than it had ever been. In 1633 Galileo Galilei, after publishing his arguments for a Copernican cosmology, with the earth and the planets revolving around the sun, rather than with the earth at the centre of the universe as had long been the prevailing view, would be tried for heresy and his works banned.

Like his masters in the Church, Gigli was a conservative man. Over the years we get a good idea of what he approved of and what he didn't. Gigli was something of a prig, and there was much that went on in Rome that made his lip curl in distaste. Despite that, he had a fine eye for daily life in the city – the storms, the fires, the earthquakes, the fate of the jailed heretics who were often hung, drawn and quartered, their various portions exposed to the populace as an example to those who might be tempted to question the Pope's authority. He wrote about the availability of bread, about miracles, comets and eclipses of the sun. Although he was a rational, educated man, there was a medieval part of Gigli that could not help but believe that God sent signs to his people below, good signs if he was pleased with them and punishments when he was enraged.

On a more down-to-earth level, Gigli wrote about people's anger at the rising taxes, their irritation at the inflexibility of the authorities and their fear of the Tiber's floods, which he always described as the river getting 'out of its bed'. In 1630 he wrote of the arrival in Rome of an elephant, 'which no one had seen for a hundred years', since the King of Portugal sent one as a gift to Pope Leo X in 1514. This particular beast, by contrast, had been brought to the city by a private citizen, and anyone who wanted to see it had to pay him one *giulio* for the pleasure.

In turn observant, witty and fastidious to the point of pernicketiness, Gigli became Rome's most wonderful portraitist. At the start of the seventeenth century, the city had a population of about 115,000. It was a tenth of the size of its imperial predecessor, smaller by far than Paris or London – smaller even than Naples – but it was growing after the ravages of the previous century, when it had been sacked and later flooded, before being visited by the plague and then cast into ruin during the papal schism. With the return of the papacy, diplomats, bankers, doctors, artisans, traders, horsemen and financiers jostled in the capital for any part of the papal chancery's swelling business.

Most of these people lived and worked, as Gigli himself did, in the *occupato*, the dense gathering of taverns and tenement houses separated by narrow cobbled streets that nestled in the crook of the arm of the Tiber where it curved in an 'S' shape, first west and then south-east. Inevitably they were almost all connected, one way or another, with the Church establishment. The three most important routes into the city converged on the Ponte Sant'Angelo, just south of the Vatican, bringing travellers who were always hungry for news and for business directly into the heart of the city. Over it rose the bell towers of nearly 350 churches. The tallest among them were the great basilicas that kept watch over the tombs of the apostles. Then there were the churches that guarded the sacred reliquaries of the saints, and occasionally, some believed, even fragments of the True Cross. Lower still were the chapels of the patrician families and the oratories of the guilds. From every sacred spire and belfry the city trumpeted its patronage of the holy. In brick and stone, every street, every façade, every arch and roof and alleyway proclaimed its thousand years as a Christian symbol.

All that lay within the *occupato*.

Beyond the *occupato*, though, lay the *disoccupato*. And that was another story altogether.

To the east of the city, starting at the Capitol, to the south and to the north ranged a barely inhabited wasteland 'set with ruins, where green snakes, black toads and winged dragons hid, whose breath poisoned the air as did the stench of rotting bodies', as an eyewitness to the epidemic that killed half of the newly crowned Holy Roman Emperor Frederick Barbarossa's army had described it in July 1155.

Encircling the town proper and extending out to the Aurelian walls, the *disoccupato* had barely changed in five hundred years. The loose patchwork of fields and vineyards was set with small houses, sheds and straw huts, tiny churches, gardens, groves and ancient ruins such as the Baths of Constantine and the Temple of Serapis. Much of it was only ever used for a few weeks each year. Dry in midwinter or at the height of a rainless summer, the *disoccupato* needed just the first downpour of spring to transform it into a swampy marsh, its muddy roads and ditches becoming pools of stagnant water that turned first green and then brown in the summer heat.

Rome at that time may have been an exciting city in which to live, but it was hardly a healthy place, though it was not until 1631 that it suffered the beginnings of the plague epidemic that would eventually kill nearly half its citizens. In an entry written at the end of December 1624, Gigli was full of apologies. 'I, Giacinto, have not been able to make daily descriptions of life as I would have liked, for I have been sick for a long time, with grave and lingering maladies, as a result of which there are many things I have not seen and others I have not noted. But, with God's pleasure, I am now well and healthy, and I hope in the Holy Year of 1625 that I will be able to make diligent note of things as they occur unlike those I have missed this year.'

Medicine had barely advanced over the centuries, and it is easy to forget how small a proportion of Europe's adult population would have been healthy at any one time. Stomach disorders of one kind or another were chronic, both among the rich, whose diet was poorly balanced, and among the poor, who found it hard to find sufficient food for themselves and their children. When they did, it was often rotten. Recurring outbursts of bacterial stomach infections resulted in dysentery, which often killed the old and the very young. Tuberculosis was rife, and for women childbirth was always very dangerous. Both sexes suffered from rotting teeth, while suppurating ulcers, eczema, scabs, running sores and other skin diseases were very common and sometimes lasted for years.

Gigli was constantly preoccupied with matters of health, his own and that of his family. And the malady he wrote about

31

most often was the Roman or marsh fever, which we now know as malaria. 'It returns every year in the summertime,' he says, 'and no one can feel himself to be safe from it.'

Rome then was the most malarious city on earth. Hundreds of people died of the disease every summer, while hundreds more were left so weak they were unable to walk, and became prey instead to the slightest infection. The rise and fall of the Tiber, which often broke its banks and flooded the plain of the Campagna, left pools of stagnant water through the country-side which provided the ideal breeding ground for the *Anopheles* mosquito that spread the disease. The views of those observers, such as the first-century BC Roman writer Marcus Terrentius Varro, who thought the miasma might be alive, full of what he called *animaletti*, 'minute animals [that are] invisible to the eye, breed there [in swamps] and, borne by the air, reach inside of the body by way of the mouth and nose, and cause disease', were regarded as extremely bizarre. Most Romans in Varro's time knew only enough to recognise the intermittent fever and shivering that visited them every year.

Giacinto Gigli had a particular reason to know about the fever. His only grandchild, Maria Cecilia Hortenzia Gigli, died of it at the age of fifteen. One day she complained of an aching head and stiff limbs. Rivulets of sweat ran down her forehead, dampening the sheets. Yet just a few hours later her mother was piling on the covers in an effort to keep the child warm. At seven o'clock in the evening, just three days after falling ill, she passed away.

Gigli was deeply affected by his granddaughter's death, and he must have fretted greatly at her decline, against which he would have had no cure other than the herbs and amulets left over from medieval times. His diary entry that day is unusually terse, and comes suddenly after a description of a great fire that destroyed the Santa Caterina de'Funari monastery. Numb with grief, he writes only that: 'She was fifteen years, five months and three days, and her beauty, her virtue and her goodness will be eternally remembered.'

The most important hospital in Rome at the time was the Santo Spirito, which had been built between the Tiber and the walls of the Vatican and which Gigli could see from his study windows. The Santo Spirito trained many of Europe's finest doctors, but for most of the city's population the cost of visiting a professional doctor was beyond their means. They preferred, in any case, to consult the herbalists and sellers of secret potions whom they had known all their lives. Many medieval cures had involved patients and physicians trying to expel their diseases by transferring them to other objects. Peasants in a number of European countries would bring a sheep into the bedroom of a fever patient, in the hope of displacing the ailment from human to beast. One cure that was still popular in the seventeenth century involved a sweet apple and an incantation to the three kings who followed the star to Bethlehem. 'Cut the apple into three parts,' advised the prescription. 'In the first part, write the words Ave Gaspari. In the second write Ave Balthasar and in the third write Ave Melchior. Then eat each segment early on three consecutive mornings, accompanied by three "Our Fathers" and three "Hail Marys" as an offering to the Holy Trinity.'

Another prescription, from a well-known sixteenth-century Roman healer named Tralliano, was supposed to be especially good against the most common fevers, called tertian and quartan because they resurged with worrying regularity, either every three days or every four. Tertian and quartan fevers were almost certainly malaria, and Tralliano's cure was the same for both: 'Take a ripe peach and remove the pip. Put the pip into an orange and tie it around the neck of the patient. He will be healed *expertum et verum.*'

Another was more complicated. 'Write the following words on a piece of paper,' it advised.

Abracolam . . .
Abracolai . . .
Abracola . . .
Abracol . . .
Abraco . . .
Abraco . . .
Abraco . . .
Abra . . .
Abr . . .
Ab . . .
A . . .

At the end, add the phrase, '*Consumatum est.*' Then have the paper tied to the neck of the patient by a young virgin using a long piece of string and reciting at the same time three 'Our Fathers' and three 'Hail Marys' in honour of the Holy Trinity.

Gigli and his fellow Romans thought they knew only too well whence spread the fever that killed his granddaughter and was as permanent a feature of the city as the smell of incense or the gentle scent of summer apricots. From the swamps and stagnating ponds of the *disoccupato*, it was believed, rose dark mists laden with fever. In Rome, went a saying, if you did not catch the fever from the *aria*, you caught it from the *mal'aria*. Bad air.

The word malaria, or mal'aria as it was always written until recently, was unknown in English until the writer Horace Walpole introduced it. In July 1740, while on a visit to the Holy City, he wrote to his friend H.S. Conway, 'There is a horrid thing called the *mal'aria* that comes to Rome every summer and kills one.' For more than a century afterwards, though, mal'aria was not taken to mean a disease so much as a noxious gas which rose from swamps or rotting carcases and vegetation, and which caused a group of ailments variously known as intermittent fever, bilious fever, congestive fever, swamp fever or ague.

Whichever of these was really malaria, the Romans had known for centuries about the miasma. From the *disoccupato* it invaded the city and forced the citizens to take to the hills every year during the worst of the summer heat, leaving the city abandoned; abandoned, that is, by those who could afford to leave. The rest stayed behind, entrusting their health to the Almighty and to the concoctions of the healers whose numbers always grew larger in summer.

Malaria had probably existed in Rome since late antiquity. Chronicles of the imperial Roman army talk of soldiers suffering from constantly recurring fever, chills, sweating and weakness, and many historians believe that one of the main causes of the collapse of the Roman Empire may well have been the prevalence of malaria around the shores of the Mediterranean Sea. In 2001, British and American scientists found malarial DNA in the bones of an infant skeleton that had been unearthed in a fifth-century villa at Lugano, near Rome.

No one is quite certain why, but malaria seems to have receded during the early Middle Ages, only to reappear with even greater severity in the years when Giacinto Gigli lived in Rome at the beginning of the seventeenth century, continuing into the eighteenth century, when it was an annual occurrence in Kent and the fenlands of England, eventually reaching as far afield as Scandinavia, Poland and Russia.

Within the Vatican, many of whose buildings were erected on Rome's lowland, by the banks of the Tiber, malaria was especially prevalent, striking with little heed for the age, rank or title of its victims. In July 1492 Bartolomeo da Bracciano, one of the senior courtiers at the palace of the Vatican, wrote to his friend Virgilio Orsini: 'The Pope, last night, had a great fever of the quartan variety, alternating between hot and cold. The Pope is confined to his bed, and it is said that perhaps he will never rise from it.' Indeed, he didn't. Four days later, on 25 July 1492, Pope Innocent VIII was dead.

Eleven years later Pope Alexander VI died, again most probably of malaria, after dining in the palatial garden of his friend Cardinal Adriano Castellesi da Corneto in August 1503. Adrian VI died of malaria in the summer of 1523, and in August 1590 Sixtus V too died of malaria at the age of sixty-nine, after a brief and very active pontificate. He had caught it a year earlier while sleeping in a hastily erected cabin during a tour of work being undertaken in the marshes around Castello Caetani, not far from Rome. Even the Borgias, who tried valiantly over the years to murder one another, could not kill each other or their enemies so regularly or so reliably as would malaria.

In the summer of 1623, shortly before Gigli, to his immense pride, was made a *caporione* for the first time, the Pope, Gregory XV, fell gravely ill. In his diary, the twenty-eight-year-old Gigli reported: 'His Holiness is not well. We must pray to the Lord.' It was said that the Pope had caught the fever the previous year, and now it had returned with a vengeance. From his study overlooking the city Gigli could see the palace of the Quirinale, nicknamed Monte Cavallo, where the Pope lay on his sickbed. An earlier Pope, also called Gregory, had chosen this superb site, less than a century before, to build his summer residence in an effort to escape the malaria that always plagued Rome during the hot summer months. In the courtyard in front of the palace, another Pope had had statues of four prancing horses installed. Nearly twenty feet high, they were Roman copies of Greek symbols of Castor and Pollux, the patrons of horsemanship who were known as the 'horse tamers', and it was they that gave the hill its nickname.

At the centre of the palace itself, dark heavy drapes shut out the light and the world beyond. For some days the Pope had been lying unmoving in his bed, covered only by a light blanket of fine wool. His head ached, his spleen was swollen and his body tormented in turn by fever and sweating, then by shivering and chills. A small troupe of Penitentiaries, the Jesuits who heard confessions in St Peter's basilica, prayed at his feet. Occasionally one would rise from his knees and another would step forward to take his place. With their gentle voices and indistinguishable cassocks of rough grey wool, they represented an unceasing rosary of care for the souls of the dying.

As a *caporione*, Gigli was often called upon to make the short journey from his home near the Via delle Botteghe Oscure to Monte Cavallo. During that long summer of 1623 he made the journey more as a way of obtaining news of the Pope's health than because there was a great deal of work to be done. For while no one knew whether the pontiff would live or die, the papal courtiers lived in an atmosphere of suspended animation, talking only in whispers. 'We are all weary,' Gigli wrote at the end of the first week of Pope Gregory's illness.

Among those who attended the Pope's sickroom was his nephew Ludovico Ludovisi. Though not yet thirty, Ludovisi had been made a cardinal by his uncle, which had enabled him to amass a considerable fortune in cash and works of art in just two years. Was his life as a man of influence about to come to an end? Should the Pope die, Ludovisi was too young to be elected pontiff himself. His only future lay in seeking to influence the choice of his uncle's successor. If a candidate with his backing should attain the throne of St Peter, Ludovisi's eminence would continue. But he had made many enemies, and would have little time to build the alliances that were essential if he were to sway the complicated negotiations that would follow Gregory's death.

As soon as the Pope died, the seal on the fisherman's ring that was the emblem of his pontificate would be broken. The new Pope would be given a new seal with his own name. Predictions of Pope Gregory's death had been made so often that he had often lamented, in the days when he felt better, that his fellow cardinals had scarcely elected him when they began planning the conclave that would select his successor. Now, it seemed, the end really had come. Gone were the badges of his office, the high, pointed, cone-shaped hat, the silken gloves. Gone too were the papal vestments with their strange names handed down through the ages – the flabellum, the falda and the fanon. On his deathbed Christ's vicar on earth wore a simple cotton shift with a wrap about his shoulders. Beneath it his pale body was only a man's, and a rotting one at that.

As Ludovisi and the other senior cardinals looked on, together with the Penitentiaries, Giacinto Gigli and the rest of the city waited outside for news. Pope Gregory's confessor began the sacrament of extreme unction. With holy oil he anointed the pontiff's eyes, his nose, his mouth, his ears. The palms of the Pope's hands had been anointed when he became a priest, so the confessor made only the sign of the cross in oil upon the backs of his hands. 'By this holy unction,' he prayed, 'and by His most tender mercy, may the Lord forgive thee whatsoever sin thou hast committed by touch.' As death drew closer, the priest began the commendation of the soul, calling: '*Subvenite*'.

In a few moments the secretary of state of the curia would knock at the door with a silver mallet, and call out for the Pope by name. Obtaining no response, he would enter the chamber and approach the bed. With another, smaller, mallet he would touch the Pope upon the forehead. Three times he would call the Pope's name and tap his cold forehead with his silver mallet. Only then would he pronounce him truly dead.

'*Subvenite*,' prayed the papal confessor once more.

'Come to his help all ye saints of God. Meet him all ye angels of God. Go forth, O Christian soul.'

It was shortly before ten o'clock at night on 8 July 1623. Pope Gregory's confessor raised his hand and with the tips of his fingers touched his head, his heart, his left side and his right. In his diary that night, Gigli wrote: '*In Nomine Patris, et Filii, et Spiritus Sancti.*'

On the night Pope Gregory died, only thirty-four members of the Sacred College of Cardinals that would elect his successor were in Rome. The other twenty or so were scattered all over the continent, some as far away as Madrid or the Baltic Sea. For a new Pope to be elected, the cardinals had no choice but to go to Rome. But the decision to travel there was not to be taken lightly. Crossing the continent, whether by sea or coach, or even on foot, was difficult and often dangerous. And Rome in the heat of summer, with the incidence of malaria rising virtually every day, was no place to be. Yet if a cardinal did not go, his vote would not be counted. He would not be able to influence the election, and as a result a Pope from a rival faction might take the throne. Knowing that Pope Gregory himself had died of the marsh fever, the cardinals who made their way towards the Holy City in the summer of 1623 did so with great trepidation. Drawing close, most of them would have elected to spend their last night well beyond the *disoccupato*, where the country air was still clear and there was little danger of breathing in the noxious gases that were believed to cause the fever. On the final day of the journey, each man made sure to rise early. The coach windows were clamped shut, and the cardinals were careful to wrap scarves about their faces, while high above the coachman would whip his horses through the approaches to the city.

That year there was trouble even before the conclave began. The interval between the death of the Pope and the election of his successor – the *sede vacante*, the vacant throne – had long been a time of release, a civic exhalation after a period of fierce papal control. By tradition, the jails were emptied. When he was *caporione*, it was Gigli's job to carry the key to the jails and oversee the prisoners' liberation. During the *sede vacante* the populace could say whatever it wanted, and the people did, many of them writing what they thought of the authorities on little pieces of paper which they then stuck on a statue of the limbless Pasquino, which is why he later came to be known as the 'talking statue'.

The papal interregnum was never so tumultuous as it was following Pope Gregory's death in 1623, when Rome erupted in an orgy of violence. It was such, Gigli recorded, as no one could remember ever having witnessed.

Not a day passed without many brawls, murders and way-layings. Men and women were often found killed in various places, many being without heads, while not a few were picked up in their plight, who had been thrown into the Tiber. Many were the houses broken into at night and sadly rifled. Doors were thrown down, women violated – some were murdered and others ravished; so also many young girls were dishonoured and carried off.

As for the *sbirri* [the papal guards], who tried to make arrests, some were killed outright, and others grievously maimed and wounded. The chief of the Trastevere region was stabbed as he went at night on the rounds of his beat, and other chiefs of the regions were many times in danger of their lives. Many of these outrages and acts of insolence were done by the soldiers who were in Rome as guards of the various lords and princes; as happened especially with those whom the Cardinal of Savoy had brought for his guard, at whose hands were killed several *sbirri* who had taken into custody a comrade of theirs. In short, from day to day, did the evil grow so much, that had the making of a new Pope been deferred as long as it once seemed likely, through the dissensions of the cardinals, there was ground to apprehend many other strange and most grievous inconveniences.

Eleven days after Pope Gregory's death, when the novena of funeral services was finally ended, fifty-five cardinals entered into the conclave to elect his successor. Three of them – Campori, a veteran of earlier conclaves, Galamina and Serra – arrived on the very evening the conclave doors were closed. Not one of them wanted to stay in the city longer than was absolutely necessary, and as it turned out they were right.

None of the French cardinals had managed to reach Rome at all, though that did not stop the envoys of the French King, Louis XIII, from seeking to influence the outcome of the election both from within and, when the papal palace doors were sealed, from without.

From the moment the doors of the Vatican were bricked up until a new Pope was elected, the cardinals lived in the papal palace, voting twice a day, morning and evening, in an effort to reach a nearly unanimous agreement on a candidate. The rest of the time, in between the obligatory attendances at mass, the cardinals lobbied and intrigued against each other, the older generation trying to hold their own against the younger men, the Spanish fighting to gain the upper hand against those supported by France or by Germany. 'We know nothing of their sacred procedures,' wrote Gigli primly. 'Nor should we.'

Of course, this wasn't strictly true. Gigli could not help but be overcome with curiosity about what was actually happening behind those sealed-up doors. By the main stove in the Sistine Chapel, he tells us, a stack of grass mixed with crushed charcoal lay ready. If, when the ballots were counted at the end of the day, no agreement had been reached, a small fire was lit. The scrutineers bound up the slips of voting paper, wet them and then burned them in the stove. The charcoal and the damp paper turned the smoke from the burning grass a dark grey, a sign to the people of Rome who stood watching that the throne of St Peter was not yet filled. Only when a new Pope was finally elected was the fire lit with grass alone, save for the last bundle of voting slips, this time dry. The smoke that curled up the chimney would be almost completely white.

With no prospect of an early agreement, the cardinals retired at night to a series of small square cubicles, cells almost, that stretched down the corridors of the Belvedere at the centre of the palace. Each room contained a narrow cot of dark wood. Hanging above it on the wall was a crucifix. There was a jug of cold water for washing, and a *prie-Dieu*. The fare was hardly luxurious. Tradition had it that if no Pope were chosen within three days, the cardinals would be restricted for five days to one dish only at supper. If after that the chair of St Peter was still vacant, they would be fed for the remainder of their stay in the conclave on nothing but dry bread, wine and water.

The tensions in Rome in the last days of July 1623 reflected those all over Europe. With the counter-Reformation, the Catholic Church was once again flexing its muscles after having been temporarily cowed by the rise of Protestantism across the continent. While it had yet to reach the extremes of the Inquisition, the Catholic power of the counter-Reformation was already a force to be reckoned with. Rome would not be so easily swept aside by the new order. In Germany, the Bohemian revolution would soon spread. France and Spain, always natural enemies, were circling each other once again. Each wanted to extend its influence over the small princeling states of northern Italy and beyond, and saw the election of a new Pope as a heaven-sent opportunity to gain the upper hand.

As always at the start of a conclave there were many interests, many candidates. There was Cardinal Sauli, who at the age of eighty-five had been a major contender in at least two earlier conclaves, and would have been so again had it not been for the fact that he was known to be completely under the influence of his valet and his wife. There was Cardinal Ginnasio, an inveterate gambler who had won 200,000 crowns in one night while he was Papal Nuncio in Madrid. There was Cardinal Campori, who had arrived at the last minute in the hope that he might this time wear the tiara that had been denied him at the previous conclave. And there was Cardinal Ascoli, a monk who regarded uncleanliness as a sign of godliness, and was generally shunned by his more urbane colleagues. There was also the dead Pope's young nephew, Ludovico Ludovisi, greedy for power and influence. Known as '*il Nipote*', it was he who introduced the word nepotism to Italian and the other Romance languages.

No clear victor emerged from the first scrutiny on the morning of 20 July. The votes of the fifty-five cardinals were distributed among several of their number, but it was already obvious that the final battle would be between two factions.

Ludovisi, despite his youth, was the leader of one group. He was hampered, however, by the fact that his uncle's short pontificate meant he had been able to create only a small number of new cardinals. The recently appointed Cardinal Richelieu, who within months would become Chief Minister to the French King Louis XIII, mentions that Ludovisi begged the Pope on his deathbed to strengthen his party with fresh nominations. This the Pope refused to do, adding somewhat unexpectedly, 'that he would already have to account to God for having made so many unworthy ones'.

The second group, which was made up largely of the cardinals who had been named by Pope Gregory's predecessor, the Borghese Pope Paul V, was more powerful. Ten months earlier, in September 1622, its leader, Scipione Borghese, Pope Paul's nephew, had given his fellow Cardinal Ludovisi a copper pendant painted by Guido Reni of the 'Virgin Sewing', but this did little to hide the fact that the two men hated one another. During Pope Gregory's pontificate, Borghese had managed to keep his faction more or less intact, even though some of the cardinals supported him with more enthusiasm than others. Yet, big as it was, this group was not strong enough to carry the day without making strategic alliances with some of the other cardinals who were supported by an array of different interests.

The French, for one, were keen to play their part in the proceedings, and Richelieu regarded the election of a francophile Pope as essential to tilting matters France's way in northern Italy, where politics were less than stable. Moreover, Richelieu knew that within the College of Cardinals was one who would be devoted to his interests.

Maffeo Barberini came from a Florentine family that had made a fortune in trade. Orphaned as a young boy, he was sent to his uncle, who was a member of the *curia*. When the lad showed promise, his uncle steered him into an ecclesiastical career, and before long he was appointed Papal Nuncio in France, where he made the acquaintance of Richelieu and the French King. This last was something of a stroke of luck. When the Nuncio in Spain, Cardinal Mellini, had been elevated to the purple, France immediately requested that as a matter of etiquette the same honour should be conferred on Barberini. Cornered, Pope Paul V, Scipione Borghese's uncle, felt he had no alternative but to comply, which he did, though with little grace. So although technically Barberini was a cardinal of Borghese's generation, he was not bound to him by any feelings of gratitude or loyalty. Richelieu, who was aware of these undercurrents, made secret arrangements with Ludovisi and the Grand Duke of Tuscany to support Barberini once their own candidates failed, as they were bound to do.

After the first day the scrutinies continued, with the voting swinging between Ludovisi's first candidate, Cardinal Bandini, and Borghese's Cardinal Mellini, a Florentine whom everyone knew would never be elected – not least because he had eighty-three nephews to provide for, which might risk carrying papal nepotism a little too far. Several days went by. The enmity between the two camps was almost physical, and Borghese and Ludovisi refused even to speak to one another. Matters were not helped by the heat, which was growing daily more oppressive. The cardinals were appalled at the idea of a protracted conclave under such unhygienic conditions.

Then, the calamity they had all feared happened. One by one, the cardinals began to fall ill with the fever. Still worse for some, more than two dozen of their attendants also became indisposed, and were incapable of attending to their duties. The cardinals' underclothes remained unwashed. Their cubicles and the passages of the Belvedere where they were housed quickly fell into a condition of nauseating neglect, 'the atmosphere being laden,' one of them wrote in his diary, 'with putrid miasmas and sickening smells of decaying victuals that the potent perfumes of the young cardinals could not manage to disguise.' As Gigli added, 'It was lacking in all dignity.'

By 3 August, after the college had been in conclave for fifteen days, at least ten of the fifty-five cardinals were ill with malaria. The next day, Borghese too succumbed. The physicians suggested potions, blistering, bleeding. Nothing worked. Borghese began thinking of leaving the conclave. All of a sudden, the francophile Cardinal Maffeo Barberini began canvassing support within his own party, supported by some of the other senior cardinals, including Ludovisi. On 5 August Cardinal Borghese had another and more severe attack of the fever. In a panic, he wrote to the Dean of the conclave asking for permission to quit the proceedings. Apprised of the fact, Ludovisi and his supporters began lobbying the Dean to refuse Borghese's request. His absence, they argued, would create a deadlock, and the entire assembly would be forced to risk their health, even their lives, for the convenience of one man.

The Cardinal Prince of Savoia was entrusted with the task of telling Borghese that the Dean refused to grant him his request. Borghese fell into a rage, and when it was suggested to him that the election of Barberini might be the quickest and simplest solution to the problem, he realised that he had been outmanoeuvred by his enemies. Judging that anything was better than running the risk of remaining in the fetid atmosphere of the Holy City, he grudgingly gave his consent.

Immediately, Ludovisi ordered the bell of the Sistine Chapel to be rung. Borghese was carried there wrapped in blankets, and Barberini's election took place at once. When the votes were counted, he fell on his knees to pray. Rising, he announced that he accepted the conclave's choice, and would take the name Urban VIII. The fire in the stove of the Sistine Chapel was lit with grass only. From its chimney rose a plume of white smoke. '*Habemus papam*,' Gigli wrote in his diary.

The name Urban, many believed, was for *Urbi et Orbi* – 'For the city and for the world' – the motto of the city of Rome over which Barberini, as Pope, would soon preside as both temporal and spiritual leader.

But the Holy City was about to demonstrate that it had powers of its own. 'As soon as they left the conclave,' wrote Giacinto Gigli, 'nearly all the cardinals fell ill and many were on the point of death. Even Pope Urban himself was among the sick.'

By the beginning of August, less than a month after Pope Gregory's death, the summer epidemic of malaria was spreading all over the city. Hundreds of people lay sick in the Santo Spirito hospital, by the Vatican. On 16 August a papal *avviso* reported that forty of the cardinals' attendants had died of the fever. One of the cardinals had already succumbed. On 19 August it was the turn of Cardinal Serra, one of those who had arrived just as the conclave doors were closing. Four days later Cardinal Sauli, who had been a possible candidate for the papacy, also died of the fever. By mid-September four more cardinals were dead, making a total of six, more than a tenth of those who had assembled for the conclave.

Outside the Vatican, the priests who said mass in the small churches on the lower reaches of the Tiber, and the lay members of the city's many confraternities who worked so diligently among the poor, died in even greater numbers.

The new Pope too could not throw off his illness. Racked with fever, alternately hot and then shivering with cold, he could feel his spleen hard and swollen by the malaria. His coronation was delayed by nearly eight weeks. Even then, he had barely recovered. At the end of his coronation day Urban's head ached. His neck was stiff, and for many weeks afterwards, one of his courtiers wrote, he could not bear the weight of the coveted papal tiara upon his head. Giulio Mancini, the senior doctor at the Santo Spirito hospital, was summoned to attend him. The new pontiff took to his bed. For nearly two months he did not leave it. Not until early in November, when the temperature had fallen and the summer fever died down, would Pope Urban be strong enough to undertake the ceremony of the *possesso*, when he would ride across Rome in a procession that saw him symbolically take possession of the Holy City. There were many who had feared that the new Pope would never be well enough to rise from his bed at all. But Urban would confound them all.

The newly-elected Pope was an educated man; yet although the early days of his pontificate were distinguished by a flourishing of the arts and the sciences, he was also deeply conservative, and in time that aspect of his character would prevail. Despite his championing of artists like Bernini and Boromini, his rule over the Roman Catholic Church would be known more for how it shackled its subjects than for how it liberated them through progress. Urban VIII imprisoned Galileo. He waged war across Europe for years at a time, financing his soldiering by imposing such high taxes on the city that he became known as *Papa Gabella*, the Tax Pope. Yet, having been educated by the Jesuits at the Collegio Romano, he also supported the quest for scientific knowledge and education that they were promoting; indeed, on the very day of his election, 6 August 1623, he issued the bulls of canonisation that made saints of Ignatius Loyola and Francis Xavier, the two men who had founded the Society of Jesus a century earlier. The Jesuits believed in educating first, converting later. Pope Urban became a great patron of Catholic missions abroad, and well before the middle of the seventeenth century there were Jesuit missions as far afield as China and South America.

A year after his coronation, Urban paid an official visit to the Santo Spirito hospital to confer a papal blessing upon Giulio Mancini and the other doctors who had helped save his life when he was sick with malaria.

From its earliest history, the order of the Confraternity of Santo Spirito had a special link with the Vatican. It was the conduit through which the Pope directed nearly all his charitable giving, and Giulio Mancini would remain Urban's personal physician throughout his reign. One of its surgeons became a specialist at dissecting and embalming. It was he who would be assigned the delicate task of embalming the Pope when he died in 1642.

The Ospedale Santo Spirito in Sassia, to give it its proper name, had the official task of caring for poor pilgrims who flocked to the city in Holy Years. An earlier Pope had built a hospice there for sick paupers after he had a dream in which an angel showed him the bodies of Rome's unwanted babies dredged up from the Tiber in fishing nets. As many as fifty wetnurses were employed in the hospital at any one time, each being able to suckle two or three babies.

The hospital Pope Urban visited could accommodate the wounded and the fevered in 150 beds, and as many as four hundred during the summer epidemics of malaria. Twice a day each doctor, accompanied by his assistant and the assistant apothecarist, would visit one of the four wards, each of which normally held about forty patients. He inspected and palpated the patients and questioned them about their symptoms. He would scrutinise their blood, which after every bloodletting was kept in a special niche by the bed, and he would prescribe treatments.

Although a special ward was reserved for the nobility, and some of the hospital's doctors also treated the cardinals and bishops who resided within the Vatican – as well as the Pope – the Santo Spirito was primarily intended to serve the poor. Most of the patients would have been artisans – blacksmiths, tailors, horsemen, bakers and butchers – but there were also many beggars who were cared for by lay nurses. Johannes Faber, a German physician who studied at the hospital, recalled that in 1600, when he began his five-year training, more than twelve thousand people received shelter, food and treatment from the Santo Spirito, as well as medication from the apothecary which had been established on the ground floor.

Under Pope Urban, the apothecary of Santo Spirito would

become one of the greatest centres for dispensing medicine in Europe. It was here that quinine, in the form of dried cinchona bark, would be given to the malaria patients in the city for the first time. In 1630 the Pope named a Spanish archbishop, Juan de Lugo, a Jesuit lawyer and university professor, as director of the apothecary. Elevated to the purple in 1642, Cardinal de Lugo would become responsible for turning the pharmacy from an artisanal studio to something approaching an industrial production line.

Like an apothecary that was being built at the same time by another Jesuit across the seas in Lima, Peru, de Lugo's Roman medicine house resembled nothing that had gone before it, either in scale or in vision. By the time Archbishop de Lugo took charge of the apothecary of the Santo Spirito hospital, its shelves were filled with recipes for preparations of medicines, prescriptions for their use and descriptions of illnesses and symptoms treated by different physicians. Spread on long tables were all the instruments of preparation: pestles, mortars, presses, beakers, alembics, boilers, distillating tubes, glass containers and ceramic jars. Neatly labelled in thousands of jars and bottles were botanical and chemical ingredients. Camillo Fanucci, one of the hospital's Jesuit apothecaries, wrote in his *Treatise on all the Pious Works of the Holy City of Rome*: 'I resolve to tell Monsignor Teseao Aldobrando, *commendatore* of this hospital, that after looking over the hospital accounts, every year we distribute more than fifty thousand syrups, ten thousand medicines and twenty-five thousand other medicines. And thus, it is obvious to anyone that no expense is spared in this hospital in the care of the sick.'

Travellers from abroad would bring small quantities of new cures to Rome. One Jesuit, travelling back from China, brought rhubarb, which would become widely used for stomach disorders. Another, from South America, came with bezoar stone, calcium phosphate that is formed in the stomach of the llama, which would become highly prized for treating all manner of ailments, from dysentery to infertility.

Yet another priest, also a Jesuit, carried back a small bundle of dried bark, the bitter-tasting outer skin of the cinchona tree, that was used by some Andean Indians of northern Peru as a cure for shivering. The priest, who knew about the marsh fever that was so prevalent in Europe, thought the powdered Peruvian bark might be worth trying against the marsh fever that struck the people of Rome during the summer, causing them repeated attacks of the sweats followed by shivering.

Thus it was that, in a prescription for curing fever noted down in the early 1630s, a Jesuit priest, Father Domenico Anda, the chief apothecarist at the Ospedale Santo Spirito, made the first passing mention of quinine – or to give it its botanical name, cinchona, which was then known as *Corticus peruvianus*, the 'Peruvian bark'.

Acc. Flor. Samb.iii
Sal.c.s.   *Cortic.Peruvian.i*
S.diapol.a   *Stib.diapol.*
Sir.giov.ii   *Sal.Tart.a a g.XV*
Spir.Theria.cum p
Fac pulverem et irrora oleo Matth. Et cum diascord. Fraest.pul.et ita per triduum.

If you go today to the Santo Spirito hospital and look around the rooms where Father Domenico had his apothecary, you see immediately how important the cinchona bark was to the development of medicine and to the reputation that the Roman apothecary would gain throughout Europe. Around the walls is a series of ceramic tablets. They show Pope Urban's Spanish priest, Cardinal de Lugo, visiting a feverish patient. At the bottom of one of the tablets is written the words: '*Purpureus Pater his solatur in aedibus aegros deluges Limae cortice febrifugio*' (In this abode, Cardinal de Lugo offered comfort to the sick with the febrifuge bark from Lima). With one hand the Cardinal crosses the patient gently on the forehead; with the other he offers him the Jesuit cure that will help drive away the Roman fever, stay the chills and ease his aching bones.

Father Domenico's prescription was referred to in a pamphlet written by Pietro De Angelis, the director of the Santo Spirito in the 1950s, who gave himself the task of educating the public about the varied work of the hospital. The original, however, no longer exists. It had been held for many years in the library of the hospital's most famous director, the seventeenth-century physician Giovanni Battista Lancisi. The library was closed to the public in the early years of the twentieth century because the building was considered unsafe. Repairing it fell foul of Italian bureaucracy and inefficiency, and it would remain closed for more than sixty years. When finally it was reopened in the mid-1990s, Father Domenico's prescription and three other of the rarest documents in Lancisi's collection had simply vanished. But a record of the text survives in Pietro De Angelis's pamphlet.

The medical world in Europe, which had barely progressed since medieval times, would take a spectacular leap forward from the 1630s with the adoption and distribution of cinchona bark in Rome. Not only was quinine the first real treatment for the Roman marsh fever, but the way it worked ran counter to the prevailing orthodoxy about fever as a disease and what was at the root of it. As a result, quinine can be described as the modern world's first real pharmaceutical drug. In time, it would change medicine forever.

That Europeans learned about it at all can be attributed to the work of a lay monk by the name of Agustino Salumbrino. A determined and energetic man with a quick, restless mind who stood not five feet high in his sandals, Brother Salumbrino had worked as an infirmarian on the wards of the Santo Spirito hospital. Unmarried, and with nothing to tie him to Rome, he set sail in 1604 for Peru, where he was determined to serve the Society of Jesus and heal the sick, and where eventually he founded the most famous pharmacy in Latin America.

The medicine he sent back to Rome came too late to treat Giacinto Gigli's young granddaughter. But for nearly a century, all the quinine that was dispensed in Europe would come from Brother Salumbrino's apothecary in Lima.

3

The Tree Discovered – Peru

> '*Aquí tenían los Jesuitas un local donde expedían al*
> *público una corteza febrífuga de la quina o cascarilla.*'
> (From this place the Jesuits provided the public with a
> febrifuge made from quinine or bark)
>
> Street plaque on the Jesuit church of
San Pedro, Lima

> *A vicuna, a cinchona tree and the horn of plenty.*
>
> The Peruvian national emblem,
as seen on every Peruvian coin

In 1663, Sebastiano Bado, a doctor from Genoa, published an account of a story he had heard from an Italian merchant who lived for many years in Lima, the capital of the viceroyalty of Peru.

The Countess of Chinchón, the wife of the Viceroy, fell ill with a tertian fever, which, Bado wrote, 'in that region is not only frequent but severe and dangerous'. Rumours of the Countess's impending death spread through the city of Lima and beyond, even reaching the Andean hill town of Loxa, in what is now southern Ecuador. On being told of the Countess's illness the Prefect of Loxa immediately wrote to her husband recommending a secret remedy he knew of, a concoction made from the bark of a local tree, which he said would cure her of all her ills. The Viceroy sent for the Prefect, who brought with him the remedy. Eagerly the Countess took it, and 'to the amazement of all', wrote Bado, 'she was cured'.

As soon as the people of Lima learned of the Countess's miraculous recovery they begged her to help them, for they had often suffered from the same fever themselves. The Countess at once agreed. Not only did she tell them what the remedy was, she ordered a large quantity of it to be sent to her so that it could be dispensed to the poor and the sick. In their gratitude the people named the cure 'the Countess's Powder'.

For more than three hundred years this sugary story was accepted as the true version of the discovery of quinine, the world's first pharmaceutical drug, that was carried back to Europe by the grateful Countess. It led to all sorts of literary fancies, most of them mercifully now forgotten. In its day the best-known was *Zuma*, written in 1817 by the Countess de Genlis, in which an Indian maid in the service of the Viceroy's household discloses the virtues of the Peruvian bark when her mistress, the Countess of Chinchón, falls ill with malaria. Other variants of this tale include *Hualma, the Peruvian*, a German novel about the discovery of quinine by a pseudonymous author, W.O. von Horn, and *The Saintly Vicereine*, a play by a Spanish poet, José María Pemán, the composer of General Franco's preferred national anthem. Written in 1939, *The Saintly Vicereine* played for a while to enthusiastic European audiences in search of an evening's distraction from the impending war, then faded quickly away.

The problem with the story of the Countess's miraculous discovery, however, is that it is completely untrue.

The Countess of Chinchón died suddenly in Cartagena on 14 January 1641, on her way back from Peru to Madrid, though her husband's diaries show she was rarely ill before that, and never with anything resembling malaria. Malaria may well have struck the Count, the Viceroy of Peru, on more than one occasion; he even seems to have suffered from it after he returned to Spain. In time he recovered, but the detailed diaries left by his secretary, Antonio Suardo, make no mention of tree barks or miraculous remedies of any description.

That the fable of the Countess's miraculous cure continued to be retold may have much to do with the fact that cinchona bark, in the early seventeenth century, really was a miracle cure. Here was an incomprehensible disease – malaria, marsh fever or the ague, as it was then called – that had been the scourge of Europe for centuries, while the cure for it was to be found high in the dense forests of a mountain range halfway across the world. The word 'malaria' did not then exist, and no one knew what really constituted agues – whether quotidian, tertian or quartan – or how people caught them, let alone how they might be cured of them. Nor, when they came eventually to learn about cinchona bark, did the doctors and apothecarists who prescribed the cure really understand how it worked either.

So how did anyone ever make the connection? How was it possible that a Jesuit priest, with little knowledge of doctoring, came to understand enough about the medicinal properties of the bitter bark to know that it might prove useful in treating malaria, a disease that would not be fully understood for another two centuries?

Some nationalistic South American historians have insisted, with little evidence, that the Spanish conquistadors must have learned about cinchona's fever-curing qualities from the Incas. While it is certainly true that the local Indians were renowned for their knowledge of plants, poisons and cures, there is scant evidence to support the argument that they knew cinchona bark cured malaria. The conquistadors wrote home about many things in the century after they first arrived in Peru in 1532, but cinchona is not mentioned by any of them. Other historians insist that the Incas kept back the secret of the miraculous fever-tree to show their displeasure at the Spanish occupation. While theoretically possible, this is unlikely, given the extent and complex nature of the contacts between the conquistadors and the local populations they encountered in South America.

The reality is that many Peruvians may not have known that the bark existed at all, at least not as a cure for malaria. The cinchona tree grew in small isolated clumps in the foothills of the high Andes. And although malaria has existed in Peru since the days of Christopher Columbus, it is found in areas of low altitude, as it is in Africa, and not at the heights where the cinchona tree grows most happily.

According to contemporary written accounts, the Indians who lived in the Andes sometimes drank infusions of cinchona bark to stop them shivering. But the observation that it might also cure marsh fever, or tertian ague, came only a century after the first conquistadors arrived in the New World, and it was made not by the local Indians, but by the European visitors.

Two Spanish writers living in Peru were the first to make any detailed description of the effects of cinchona bark on patients suffering from the ague. In 1638 an Augustinian friar and herbalist, Antonio de la Calancha, wrote: 'A tree grows in the country of Loxa which they call of fevers, whose bark, of cinnamon colour, made into powder given to the weight of two *reals* of silver in a drink, cures the ague and tertians; it has produced in Lima miraculous results.'

Calancha had been born in Chiquisaca (now Sucre), in the highlands of Bolivia, in 1584. He grew up among the Andean Indians, and was intimately aware of their customs and folk medicine. He entered the Augustinian Order in 1598, and was appointed the Rector of St Idelfonso College in Lima nearly twenty-five years later. Calancha spent much of his adult life writing his nine-hundred-page *Corónica moralizada de la orden de N.S.P.S. Agustín en el Peru*, and his account of the properties of the cinchona bark was probably written around 1630, the year that the Viceroy, the Count of Chinchón, first fell ill with the ague, as noted in the diary of his secretary, Antonio Suardo.

Another priest, Bernabé Cobó, a Jesuit who arrived in Lima from Spain in 1599, wrote an account of cinchona as a short chapter entitled 'A Tree for the Ague' in his magnificent multi-volume *Historia del Nuevo Mundo*, which was written in 1639 but not widely disseminated for another two centuries. In it he says: 'In the district of the city of Loxa, diocese of Quito, grow certain kind of large trees, which have bark like cinnamon, a bit coarse and very bitter; which, ground to powder, is given to those who have the ague and with only this remedy it is gone. These powders must be taken to the weight of two *reals* of silver in wine or any other liquor just before the chill starts. These powders are by now so well known and esteemed, not only in all the Indies, but in Europe, that with insistence they are sent for from Rome.'

The writings of Calancha and Cobó were well known to virtually everyone who has written about cinchona or quinine over the past hundred years. Four other Spanish writers, all of them far more obscure, bear out Calancha and Cobó's observations that cinchona came to the attention of the Jesuits in Peru in or around 1630.

Gaspar Caldera de la Heredia was born in Seville of Portuguese extraction in 1591 or thereabouts. He studied medicine at the University of Salamanca, practising first in Carmona before settling in Seville, then already the centre of Spanish imports from the Indies. Caldera's interest in remedies from the New World is easily understood; his father had lived in Mexico, and three of his children went to Lima in 1641, precisely at the time when the use of cinchona in Spain and other parts of Europe was gaining momentum. His writings on cinchona are preceded by a series of letters that he exchanged in 1661 with Girolamo Bardo, the pharmacist at the Jesuit College in Rome and a close collaborator of the doctors from the Santo Spirito hospital who cured Pope Urban VIII of the malaria he caught during the papal conclave that elected him.

Caldera's *Tribunalis Illustrationes et Observationes Practicae*, in which he writes about cinchona, was published in 1663, the same year that the Genovese doctor, Sebastiano Bado, published his celebrated book on cinchona, *Anastasis Corticis Peruviae Sen Chinae Chinae Defensio*. Caldera's writing shows him to be a learned man, a cautious scientist, a sound clinical practitioner and a faithful witness. Cinchona, he wrote, came from a tree like a large pear tree called *quarango* by the Indians, who used it as timber. Jesuits at missions in the foothills of the Andes noticed that the Indians drank its powdered bark in hot water when shivering after being exposed to dampness and cold. Quinine has many side effects, some of them quite unpleasant, such as tinnitus, but one of its more beneficial properties is that it can act as a muscle relaxant, which is why it calms the nervous impulse that causes shivering, and why today it is sometimes prescribed for people with pacemakers, or more commonly for those who suffer from leg cramps.

The Jesuits, Caldera noted, believed that cinchona might be effective in checking the shivering that is associated with ague, and they tested the powdered bark on a few patients suffering from quartan and tertian fever. Shortly afterwards, some Jesuits of the missions in Quito took the bark to Gabriel de España, an energetic pharmacist who had his *botíca* in Lima near the bridge over the Río Rimac, and who was renowned throughout the young city for his knowledge of local medicinal plants. De España began to pass samples of cinchona to a number of physicians as well as other apothecaries in the city, who used the bark in the treatment of intermittent fevers with great success.

Gaspar Bravo de Sobremonte, who studied medicine at the University of Valladolid, where he held several chairs including Surgery, Method and Medicine, also wrote about cinchona. Bravo was considered one of the best physicians of his day, and many of his works were published in Spain and France. In the second edition of his *Disputatio Apologetica pro Dogmatica Medicine Praestantia*, which was published in 1639, he describes how the Spaniards – 'us', he calls them – used Peruvian bark to treat intermittent fevers after observing Indians in Peru drinking the powdered bark in hot water when they were shivering with cold.

In the 1670s, two other Spanish doctors also wrote about the curative effects of cinchona. Pedro Miguel de Heredia (no relation of Gaspar Caldera de la Heredia) studied medicine at one of the greatest of the Spanish faculties, Alcalá de Henares. There he held the chair of *Prima* of medicine, retaining it several times after the compulsory contests that took place every four years. Miguel left Alcalá de Henares in 1643. More than forty years later, the second edition of his four-volume

Operum Medicinalium recounted how the Jesuits in Peru had tested cinchona.

Similarly, Miguel Salado Garcés, who held the chair of Method at the University of Seville and was committed to discovering every new drug that came from America, wrote in 1655 in his *Estaciones medicas* that 'the missionaries of the Society of Jesus [in the province of Quito] used the powders of Quarango following the second transit of Galen with great ingenuity, after observing that the Indians took them when shivering from cold after swimming in iced water or from the coldness of the snow, and stop trembling within a short time; [the Jesuits] used them to control the shivering in tertian and quartan fevers: and as they noticed that the repetition of the fever stops, they advised them as a great febrifuge (and they still continue to do so) to cure them . . .'

Caldera, Bravo, Miguel and Salado Garcés all put the Jesuits at the centre of the story of the early discovery of cinchona in Peru. But who was responsible for gaining it such wide renown in Europe?

In the spring of 1605 a small group of Jesuit priests disembarked at Callao, at the mouth of the Río Rimac downstream from Lima. For nearly three hundred days they had been tossed about, never knowing a moment's quiet as they rode the swells of the vast Atlantic Ocean on their journey towards the southern tip of South America. The final part of the voyage, hugging the continent's west coast, was if anything worse than the open sea. Gigantic waves hurled themselves across the vessel, throwing up thick columns of spray that then collapsed upon the deck, drenching everything in a foamy swirl and threatening to drive the ship onto the jagged rocks.

Now that they had reached dry land, their leader Father

Diego de Torres Bollo urged them ashore. As he called for yet more donkeys to carry the supplies that the priests had brought with them from the Holy City, one of their number, a small man in sandals and the rough brown tunic of a lay brother, broke away to look around him.

Agustino Salumbrino was then about twenty-five years old, but his beard was thick and he looked older, for he had started work when he was still very young, and had never taken a single day of rest since. While Salumbrino had already studied and travelled more than most men would have done in a lifetime, he knew that Peru was the place where he would spend the rest of his days. More than that, he knew exactly what he would do with his time there.

Francisco Pizarro's conquest of Peru in the 1530s was well known to Europeans by the beginning of the seventeenth century, for at least two of the conquistadors who had travelled with him to South America had written widely-read accounts of the magnificent Inca civilisation. Pizarro's conquest was driven entirely by greed for Inca gold and treasure, but he painted it with a religious sheen to give legitimacy to his actions. Accompanying him on his first journey to the New World was a troop of Dominican priests. Four decades later, on 1 April 1568, the first Jesuit priests, eight of them in all, arrived in Lima.

The city was then only thirty-three years old, and still known by the name Pizarro had given it, *La Ciudad de los Reyes*. Despite its strange microclimate, which casts a thick fog over the coast for nearly ten months of the year, the City of Kings deeply impressed the small party of Jesuits. They admired the formal chequer pattern of the streets, so characteristic of sixteenth-century Spanish towns, that extended in a straight line right down to the Río Rimac, which now runs through the centre of the city. They thought highly of the beautiful public square in front of the viceregal palace, the monasteries of the religious communities and the buildings of the civil and ecclesiastical administrators. Father Diego de Bracamonte, one of the newly arrived Jesuits, paid Lima a handsome compliment when he wrote home describing the city as 'another Seville'.

Among the first tasks to which the young Jesuits set their minds was finding a suitable location for housing the fledgling mission. Before paying a call on the acting Governor, Lope García de Castro, they explored the city. They soon chose a square, three blocks to the east of what is now the Plaza Mayor, which then housed the Viceroy's palace, and three blocks from the Franciscans, in a rather densely populated area of the city. After a brief public hearing, the Jesuits were granted the expropriated property, for which they were obliged to pay twelve thousand pesos in compensation to its former owners. The transaction, which was completed in just over two weeks, makes it sound as if the Jesuits had arrived in Lima with plenty of money. If anything, the reverse was true: for thirteen years after they moved in to San Pablo the Jesuits had to assign one of their most capable lay brothers to be the *limosnero*, the man in charge of begging alms on a daily basis from the well-to-do families of the city.

In the early years, the Jesuit College of San Pablo depended for its existence on these donations from the citizens of Lima, and a series of small and sporadic royal grants. In 1581, though, San Pablo took over a property outside the city, and over nearly two centuries, until the Jesuits' expulsion from the Spanish Empire in 1767, those holdings steadily increased in size until the Society of Jesus became one of the country's biggest landholders. Its haciendas produced wheat, which was ground into flour in a mill that was also owned by the college. The Society planted new vines and an olive grove, which provided the Jesuit fathers with as much wine and oil as they needed. They raised cattle and goats, and grew sugar cane. A *trapiche*, or sugar mill, produced sugar and cane syrup. By 1600 San Pablo owned about ten rural properties, of which some were put under intense cultivation while others were used for grazing.

The haciendas fed and clothed all the 160 or so priests who lived at the college in Lima. By the first half of the seventeenth century they also supported two thousand workers whom the Jesuits employed to run their properties, and three hundred slaves who were engaged in the vineyards of San Xavier, picking and pressing the grapes and producing the well-known Jesuit wines, as well as *pisco*, the traditional Peruvian liquor that is distilled from white grapes. As the haciendas grew bigger and more efficient, they turned from being simple agricultural properties into agro-industrial plants – a fusion of farms with mills, sugar refineries and distilleries – which delivered to Peruvian markets some of the best wines, flowers, sugar, oil and honey available in the viceroyalty.

Over the years many Jesuits sailed across the Atlantic to join the missions that were being set up in Chile and Argentina, as well as Peru, but there was always room for more. In order to expand throughout the viceroyalty, the Jesuit mission in Lima had to have more resources. And that meant more people.

Thus it was that in 1601 Diego de Torres Bollo, one of the senior Jesuit priests at the mission of San Pablo in Lima, left for Rome to petition the Vicar-General of the Order of the Society of Jesus to send more young Jesuits to South America. To reach the Holy City he had had to sail around the north-west coast of South America to Panama, then travel by mule over the isthmus to Puertobelo before resuming his journey once again by ship. The voyage took many months, and was fraught with danger. Not long after he arrived, Torres Bollo fell ill and was admitted to the Jesuit infirmary in Rome. The man who took care of him was Agustino Salumbrino.

Salumbrino had joined the Jesuits in 1588. After taking his vows in Rome in 1590, he was sent to the Jesuit college in Milan to become an infirmarian. There he made a special study of pharmacy, and when, after a few years, he returned to Rome, he resolved to put his medical knowledge at the service of the many Jesuits who lived in the Holy City. In the course of his convalescence, Torres Bollo, who would later found the Jesuit mission in Paraguay, told Salumbrino all about St Ignatius' missions in the New World, and the great college of San Pablo which was being built in the young city of Lima. He described the plans the Peruvian Jesuits had for setting up new missions over the whole continent.

Each time he came to see his patient, Salumbrino, like the Mill Hill fathers whom I had visited in north London, felt the call of the mysterious world across the seas. As he listened to Torres Bollo, the lay brother began to recognise what his life's work would be. He would go to Peru and live in the college of San Pablo in Lima, putting his knowledge of pharmacy to good use as he built up the *botica* into the best pharmacy in the New World, which for nearly forty years, until his death in 1642, is exactly what he did.

Today, the soaring Baroque church of San Pedro, next to the Biblioteca Nacional in the centre of Lima, is all that is left of the great Jesuit College of San Pablo, which rapidly fell into disrepair after the Jesuits were expelled from the Viceroyalty of Peru and the rest of the Spanish Empire in 1767.

The church is dark, though well cared for. The remnants of its first crucifix are housed in a glass cabinet in a corner. The dark wood gleams in the shadows. In another corner, a vast reliquary rises. When you look at it more closely, you see that it is made of hundreds of boxes of dusty human bones, said to come from the many local priests who have been canonised and then forgotten.

The priest who says mass there today, under the fifty-two crystal chandeliers strung along the ceiling, urges his congregation, as Catholic priests do everywhere, to 'go in peace, and to love and serve the Lord'. Despite the medieval, talismanic quality of the church's interior, its enormous congregations reflect the power that the Catholic Church still commands in Peru, and the extent to which it has permeated every level of political and intellectual life. Alejandro Toledo, who won the presidential election in 2001, considered including two Jesuits in his cabinet. One was a well-known figure, Father Juan Julio Witch, an economist and academic who became a household name in 1997 after the Japanese embassy in Lima was taken over by terrorists during a reception to celebrate the Emperor's birthday. Given the opportunity to leave the embassy compound, Father Juan elected instead to stay behind with the other hostages. For four months he said mass there every day, urging his small and frightened congregation to keep their spirits up and pray for salvation, while outside television cameras and frustrated army marksmen waited for the terrorists to give way. Today his name is known throughout Peru.

It has long been accepted that the Jesuits were responsible for introducing cinchona into Europe. As I read more about the role of the Society of Jesus in Peru's long history, I kept wondering if, after all the wars and insurrections, burning and looting, any written material still survived from their earliest time there. Eventually, I learned that the bulk of a large private collection of Peruvian Jesuitica, built up by an obsessive hoarder, Father Rubén Vargas Ugarte, who came from a wealthy family and was himself a Jesuit priest, a historian of the Church and a Peruvian nationalist, had been placed in the state archives in Lima. No one could tell me over the telephone what the collection contained. And so I travelled to Peru, not knowing what I might find, or indeed whether I would find anything at all.

For a long time after I got there, I was little the wiser.

The state archive occupies a dark corner at the back of the building that is better known as the Peruvian national bank. It has its own entrance, but there is no sign on the door; nor, once you are inside, are there any directions to tell you where to go. At least Father Rubén's material was there, though what it contained no one could tell me. It hadn't been opened, and although some of its papers had been looked over by scholars past, the collection had come with no inventory and there was no catalogue. But I was welcome, the librarian said, to look through it if I wanted. A long pile of boxes was stacked down the hallway, though quite how far it extended and how many boxes it contained I couldn't really tell.

The librarian gave me a key to the archive, and said I could work there for as long as I wanted each day. Outside, Peruvians were preparing to go to the polls. The atmosphere was tense, and there was a sporadic curfew. Occasionally I would emerge from the archive at the end of the day and find that I could not return to my hotel. I went back inside and slept on the floor, shielded from the draught by Father Rubén's mountains of paper.

I began methodically going through each box. I picked out details of property transactions, plantings and harvests on the Society's haciendas, chronicles of boundary disputes, baptismal records, the sale and purchase of slaves, shopping lists, inventories. I had come across Agustino Salumbrino's name in the Jesuit archive in Rome, but I still knew very little about him. Would I ever find out more?

Then one day I found two old books dating back to 1624. They were inventories. Page after page in the volumes of *El Libro de Viáticos y Almacén* are filled with inky lines that once were black but now are faded to a rusty brown. The quills the writers used were so sharp they have ripped through the paper. That these documents survived at all was a miracle. No one has ever tried to conserve them, and some of the paper crumbled in my hand. Yet it was still possible to read what the Jesuit administrator of San Pablo had written there nearly four hundred years earlier. In addition to listing everything that came into and went out of the college, from the cases of books that were sent from Europe to supplies of medicines, clothing and tableware that were despatched to other Jesuit missions, he also provided a complete inventory of Brother Salumbrino's pharmacy.

By the time Agustino Salumbrino arrived in Lima in 1605, the mission at San Pablo was firmly established, with several classrooms, a small library, a chapel, private rooms and even an infirmary, which, although basic, was clean and well run. Salumbrino quickly realised, however, that the infirmary was not sufficient to take proper care of all the sick in Lima. What he needed was a proper pharmacy, and a steady supply of medicines. These were hard to obtain in colonial Peru, but Salumbrino was a tireless worker, and he had made up his mind not merely to build a pharmacy, a *botíca*, as it was called, at San Pablo, but to do it in the grand manner, not only to serve the college's local needs, but to supply all the Jesuit colleges throughout the viceroyalty.

Agustino Salumbrino's ambition to set up a pharmacy to help treat the poor of Lima had its roots not just in the rich medical lore that he encountered as soon as he arrived in Lima, but also in the Jesuits' earliest philosophy. The instructions left by the founder of the order, St Ignatius Loyola, forbade his followers to become doctors. The task that lay before them, he emphasised, should focus upon men's souls. This did not mean that Jesuits were ignorant of the importance of maintaining good health; indeed, every Jesuit mission was enjoined to appoint one of their number as a 'prefect of health' to ensure that the priests' diet was adequate and that they were well cared for. The most capable lay brothers would be chosen to run the college's infirmary and have immediate care of the sick. Most important, the Society's founder insisted, each college would ensure that it had an adequate supply of medicines, either by setting up a pharmacy of its own, or by finding a reliable source of supply. Despite being expressly forbidden to practise medicine, Jesuit priests often turned their attention to the study of herbs and plants, and a number of them, especially in the foreign missions, became apothecaries.

San Pablo's infirmary was in a clean and quiet courtyard in the south-eastern corner of the college. By the time it was properly established it had about fifteen private rooms, all facing the fountain in the centre of the courtyard. Brother Salumbrino built his pharmacy close by the infirmary. Knowing that he needed to be as self-sufficient as possible, he began by planting a small herbarium in a corner of the garden at San Pablo. He chose plants that were well known for their medicinal properties: camphor, rue, *nicotiana*, saffron and *caña fistula*, a Peruvian wild cane that was often used for stomach disorders in place of rhubarb. These Brother Salumbrino and his two assistants made up into medicinal compounds, which were dried, powdered and mixed in the laboratory according to strict pharmaceutical rules. To help him, Salumbrino ordered two of the most important pharmacopoeias then available in Europe: Luis de Olviedo's *Methodo de la Colección y Reposición de las Medicinas Simples y de su Corrección y Preparación* (printed in Madrid in 1581), which he had used in Rome; eventually, he also ordered Juan del Castillo's *Pharmacopoea Universa Medicamenta in Officinis Pharmaceuticis Usitata Complectens et Explicans* (printed in Cadiz in 1622).

Pharmacopoeias were works that described chemicals, drugs and medicinal preparations. They were issued regularly with the approval of different medical authorities, and were considered standard manuals in every pharmacy in Europe. Besides these two classics, Brother Salumbrino could also consult and follow the prescriptions of Girolamo Mercuriale, physician to the Medicis and Professor at the universities of Padua, Bologna and Pisa, who exercised a profound influence on medical circles all over Europe.

Over the next century and a half the *botica* at San Pablo would order at least ten other pharmacopoeias that specialised in local drugs and chemicals, including its *vade mecum*, Felix Palacios' *Palestra Farmaceutica*, which was printed in Madrid in 1713, the year after Francesco Torti had his 'Tree of Fevers' published. The *botica* put in regular orders for extra copies of Palacios' work to be sent out to the other Jesuit colleges in the viceroyalty. The book was so highly regarded, and was so frequently referred to, that the pharmacists at San Pablo would eventually write inside the cover of their own copy: *para el uso diario de esta botíca* ('for the daily use of this pharmacy').

By 1767, when the Jesuits were forced to leave Peru and the final inventory of the pharmacy was compiled, the San Pablo medical library contained about a hundred books. The full list, given in another set of books that I found among Father Rubén's boxes, *Cuenta de la Botíca 1757–1767*, included the ancient classics by Galen and Hippocrates as well as voluminous Latin commentaries on the two masters by several medieval doctors. The library also had books on several other branches of medicine, including anatomy and osteology, treatises on different kinds of fevers and their remedies, descriptions of contagious diseases and their infections, and the methods of combating them.

Surgery was also a favourite subject at San Pablo, and one could find on the shelves of the college's library Bartolomé Hidalgo de Agüero's *Thesoro de la Verdadera Cirugía y Via Particular contra la Común* (printed in Seville in 1624), and Juan Calvo's *Primera y Segunda Parte de la Cirugía Universal y Particular del Cuerpo Humano*, which was published in Madrid in 1626 and reprinted many times in the seventeenth century, and was still in use more than a hundred years later – though one shudders to think of operations being carried out without the benefit of any anaesthetic or antibiotics in the humid atmosphere of seventeenth-century South America.

The Jesuits who came to Peru just after Pizarro's conquest of the Incas were the first order with a clear mission to educate and then, by doing so, to convert the Indians to Catholicism.

There was a clear division, though, over exactly how this should be done. The ascetic, intellectual Jesuits who ran the order from Rome were of one view, while the energetic activists who left their homes to promote its interests overseas were of quite another. The young Jesuits in Lima were pioneers of the soul. They believed strongly in catechism. Each day, a group of priests would leave San Pablo, walking in procession through the streets of Lima, holding a crucifix and ringing a bell to attract groups of Indians and blacks to whom they would preach. Not everyone liked this. One early Provincial Superior of the Jesuits in Peru, José de Acosta, was dismayed by this helter-skelter missionary activity, and was bitterly critical of having so many men tramp around as 'holy vagabonds'. His own bias tended towards the old Jesuit ideal of learned men influencing councils and kings.

What he failed to realise was that most of the Jesuits, such as Brother Agustino Salumbrino, who went to Peru early on were not driven to write books or meditate. They were zealous, educated men, full of drive and courage, who wanted to make a difference, whether it was by saving souls or promoting good health.

The early Jesuits soon expanded their missionary activity to Cuzco, the old Inca capital that Pizarro seized in 1534. They bought a fine palace that had been taken over by one of Pizarro's lieutenants on the main square, only to tear it down and build the towering pink Baroque church that still stands today. From there the Society sent its priests out into the countryside to make contact with the local Indian communities and urge them to renounce the animist gods they had worshipped for centuries in favour of a Christian Almighty.

The Inca world was ruled by spirits and superstition. Every village was surrounded by secret places – trees, rocks, springs and caves – that had a magical significance. The Incas collected unusual objects, and in every house there were *canopas*, or household deities, displayed in a niche in a corner or stowed in a special place, wrapped in cloths. They observed rituals throughout their daily lives, sprinkling *chicha* or *coca* when ploughing, saying prayers and incantations when crossing rivers, making sacrifices on particular occasions and always leaving an object on the pile of stones that is still often to be found at the top of every pass.

The Incas lived in fear of the sorcerers, the old men who foretold the future by studying the shape of ears of corn, the entrails of animals or the movement of the clouds, and were terrified of the magic spells they cast to cause love or grief in their victims. But they also revered them, for many of the sorcerers were medicine men as well as magicians. In some parts of Peru they would undertake trepanning, cutting open the skull to let out evil spirits and to offer the patient some relief from pain or swelling. The rich Quechua language shows that the Incas had a fine knowledge of anatomy and medicine, with words such as *hicsa* for abdomen, *cunca oncoy* for angina, *susuncay* for putting to sleep, *siqui tullu* for coccyx, *husputay* for haemorrhage, *hanqqu* for nerve and *rupphapacuy* for fever.

They amassed a great store of knowledge about local plants and how to use them to treat different ailments, and were particularly expert on poisons and plants with hallucinogenic qualities – every man would carry upon him a little packet of coca leaves for chewing on. They also used the trumpet-shaped *Solanaceae*, or datura as it is better known, in magic spells to cast their enemies into a trance, sarsaparilla as a diuretic, tembladera (*Equisetum bogotense*) against pyorrhoea, a plant they called *llaquellaque* (*Rumex cuneifolius*) as a purgative of the blood, *ortiga* (*Urtica magellanica*) to cure sciatica, and *payco* (*Chenopodium ambrosoides*) against worms.

The two volumes of *El Libro de Viáticos y Almacén* show just how elaborately Brother Salumbrino and his fellow Jesuit priests would prepare for a trip out of the city. Every traveller would be issued with a mule for riding on, and another for carrying their supplies. Many of the mules' names survive in the records: La Cabezuda, La Caminante, La Mulata, El Galán. The supplies would include hay for the mules, for the desert of northern Peru in particular was short of fodder, and often of water too. The traveller would also be equipped with a bowl, a spoon for the table and a knife for cutting meat, a bedroll and a sheet, a roll of sealing wax, spices in the form of saffron, pepper and cinnamon, wine, a sombrero, a soutane and a cape to keep out the cold in the mountains. The grandest inventories included travelling altars, supplies of wine and wheat hosts for offering communion, and even silver candlesticks. But, grand or simple, each traveller's list concludes with *patacones*, fried plantain chips, for an Indian guide, and more *patacones* for *el gasto del camino*, the road toll.

Despite the rips in the pages of these ancient books, they still summon, nearly three centuries on, a pervasive and enormously fierce sense of just how energetic and enterprising the Jesuits were. On 26 April 1628, the earliest entry in the book that mentions Brother Salumbrino, the pharmacist sent the Jesuit college at Arequipa, at least three weeks' ride south of Lima, not far from Lake Titicaca, four cases of drugs, including eight *libras* of *caña fístula*. The following month he sent the college another eight *libras* of *caña fístula* and a copy of the *Meditations* of St Ignatius Loyola. In August of that same year he despatched supplies of tobacco and cocoa and another three boxes of *caña fístula*, and the following April the mule load to Arequipa would include four bottles with different drugs 'sent by Brother Agustín'.

San Pablo was making a name for itself as a trading post, and it was not confined to medicines. It imported textiles from England, Spain, France and the Low Countries, Italy and the Philippines, and large quantities of black taffeta from China. It provided Jesuit schools in the region with ink and paper imported from Italy – in 1629 San Pablo despatched three thousand pens in a single huge shipment that went to the Jesuit College in Santiago, Chile. Farm tools such as ploughs, sickles and hoes were in great demand. San Pablo shipped those off too, along with saddles and harnesses, tallow candles and pottery, shoes and clothing for children as well as adults, needles and nails. In 1628 the college sent twelve *baras* of tailors' needles from France to Arequipa, while three years later another two thousand needles, described as *finas de Sevilla*, were needed. Between 1628 and 1629 San Pablo also sent twelve thousand nails to Potosí, ten thousand to Arequipa, and more than twenty thousand to Chile.

As this trade blossomed, Brother Salumbrino's influence also soon extended beyond the walls of the college. Like the library at San Pablo, which ordered books from Europe and sent them out to colleges all over the viceroyalty, the pharmacy became an early distribution centre of medicines and medical information for other Jesuit institutions in the area. Salumbrino supplied medicines to the Jesuits who left San Pablo on long missions among the Indians in the Andes, and to other Jesuit outposts

The *Libro de la Botíca* neatly lists everything that San Pablo supplied to the other Jesuit colleges in the viceroyalty: *agua fuerte* and *aguardiente*, powdered mother of pearl, pine resins, black and white balsam, bezoar stone, *nicotiana* in powder, *caña fistula*, cinnamon, nutmeg, sal volatile – the original smelling salts – *mercurio dulce* or mercury sulphide for treat-

My French grandmother,
Giselle Bunau-Varilla,
shortly before she eloped
to Africa in 1928.

My Italian grandfather, Mario
Rocco, in Kenya with my father
as a baby in 1930, just weeks
before he fell ill with malaria
for the first time. Raised in the
Roman Campagna, my grand-
father knew all about the marsh
fever to be caught from the
'bad air', or *mal'aria*. Without
quinine, he would fall sick
several times with the disease
while detained in Kenya and
South Africa during the Second
World War.

On a visit to the malarial province of Zeeland in 1520, the artist Albrecht Dürer fell ill with a swollen spleen, a characteristic symptom of malaria; he sent this drawing back to his doctor, writing: 'Where this yellow spot is to which my finger points, there it is that I feel pain.'

One of the four wards of the Santo Spirito Hospital in Rome that in the summer months would fill with patients suffering from the Roman or marsh fever.

Ernest Herbert's languid Victorian painting *Malaria* perfectly captures how feeble and listless patients feel when they succumb to the disease.

PERLATA COLLECTA NOVIS CHINCONIUS ORIS
ACCIPIT A SERVO PHARMACA FEBRIFUGA

AEGROTAT LIMAE CONIUX CHINCONIA FEBRIM
CORTICE MIRANDO POCULA TINCTA FUGANT

PURPUREUS PATER HIS SOLATUR IN AEDIBUS AEGROS
DELUGUS LIMAE CORTICE FEBRIFUGO

Three wall-mounted plaques in the former apothecary of the Santo Spirito
Hospital in Rome tell the story of *(top)* how quinine was brought to the
attention of Jesuit priests by the Prefect of Loxa, in southern Ecuador;
(centre) how it was introduced to Europe by the Countess of Chinchón;
and *(bottom)* how it was distributed to the poor of Rome by the
Spanish cardinal, Juan de Lugo

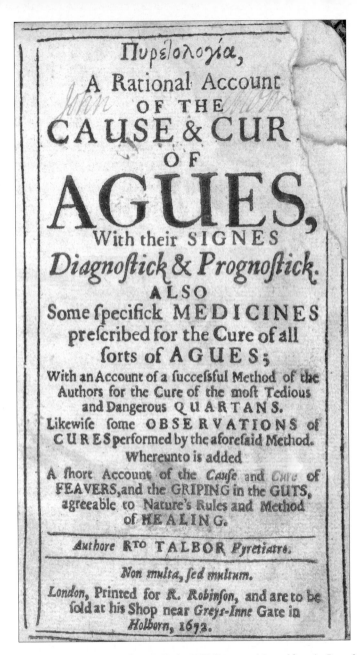

Πυρετολογία,

A Rational Account
OF THE
CAUSE & CUR
OF
AGUES,
With their SIGNES
Diagnostick & Prognostick.
ALSO
Some specifick MEDICINES
prescribed for the Cure of all
sorts of AGUES;
With an Account of a successful Method of the
Authors for the Cure of the most Tedious
and Dangerous QUARTANS.
Likewise some OBSERVATIONS of
CURES performed by the aforesaid Method.
Whereunto is added
A short Account of the *Cause* and *Cure* of
FEAVERS, and the GRIPING in the GUTS,
agreeable to Nature's Rules and Method
of HEALING.

Authore R^TO TALBOR *Pyretiatro.*

Non multa, sed multum.
London, Printed for *R. Robinson,* and are to be
sold at his Shop near *Greys-Inne* Gate in
Holborn, 1672.

The cheeky Cambridge doctor Robert Talbor set himself up in London
as a *pyretriato,* a fever specialist. The secret fever cure that brought him to
the attention of King Charles II of England and King Louis XIV of France
was found, after his death, to be largely made up of quinine.

ing syphilis, black pepper, ambergris, senna, tamarind, sugar, camphor, sweet and bitter almonds, almond oil, tobacco from Seville, essence of roses and violets, rhubarb, chocolate and, of course, cinchona bark, that would eventually be despatched, dried in strips or in powder, in huge quantities all over the continent and also across the Atlantic.

From the earliest years the Jesuits of San Pablo were of the clear belief that conversion of the Indians would come about not by force, but by education and persuasion. For that reason they were quick to send young priests out into the field. Many of the young Jesuits who were posted to Peru made it a priority to learn Quechua and the other Indian languages, and to accustom themselves to the Indians' way of life.

The Jesuits in the field, especially those who had been sent north-east of Lima, to Loxa in the Andes, began to persuade the local Indians to seek out the *árbol de las calenturas*, the 'tree of barks', as Bernabé Cobó, another Jesuit and a colleague of Salumbrino's, would describe cinchona in his *Historia del Nuevo Mundo* in 1639. They taught them how to cut off the bark in vertical strips so as not to kill the tree, and to plant five new trees for every one they cut down. The Jesuits would place the saplings in the ground in the shape of a cross, in the belief that God would then help them grow better. More than two centuries later, an English plant-hunter and bark-trader would observe: 'Always when passing [these plantations] my Indians would go down on their knees, hat in hand, cross themselves, [and] say a prayer for the souls of the *Buenos padres.*'

After they stripped off the bark, the *cascarilleros* or bark-hunters would cut it into pieces and leave it to dry in the sun. Taking care not to break the fragile, powdery strips, they would wrap them carefully in pieces of cloth and then in watertight leather packs for transporting down the hills by mule to Lima.

San Pablo began to distribute cinchona bark – or *cascarilla* as it was known in Spanish – to the other Jesuit colleges in the viceroyalty, and even as far as Panama and Chile. Eventually Brother Salumbrino also began sending supplies of cinchona to Europe.

The first person listed in the *Libro de Viáticos y Almacén* as leaving San Pablo with a quantity of cinchona bound for Europe is a Father Alonso Messia Venegás, an elderly Jesuit priest who carried a small supply of it in his bags when he travelled to Rome in 1631. Father Alonso knew, as every Jesuit did, how malarious the Holy City was, and had heard accounts of the terrible conclave of 1623 when so many of the visiting cardinals died. Rome was in dire need of a cure for the fevers, and Brother Salumbrino was eager to see if the plant that stopped people from shivering could be put to use curing the chills that were a symptom of the marsh fever. Little did he know that not only did it stop the shivering, it could also be used to treat the disease.

The physicians in Rome found that the bark was indeed an effective treatment for the intermittent fever, and thereafter every Procurator who left San Pablo for the Holy City to represent the Peruvian Jesuits at the congress that elected the Jesuit Vicar-General every three years would take with him new supplies of the febrifuge bark. Shortly after Father Bartolomé Tafur, who served as the Peruvian representative at the congress of 1649, arrived in Rome he renewed his acquaintanceship with Cardinal Juan de Lugo, who was then in charge of the apothecary at the Santo Spirito hospital, and was becoming cinchona's champion in the Holy City. In 1667 Felipe de Paz took with him a trunk filled with the *corteza de la calenturas*, and in 1669 Nicolás de Miravál arrived with 635 *libras* of cinchona for distribution in the *curia*, having left a similar amount in Spain.

By the second half of the seventeenth century, according to an early map of Lima in the state archive, the citizens of the capital had begun calling the street in front of the Jesuit infirmaries *Calle de la cascarilla*, Bark Street. Now part of the long, fume-laden Jirón Azangaro, which runs through downtown Lima from the Palacio de la Justicia as far as the Franciscan convent near the river, *Calle de la cascarilla* would remain up to the start of the republican period in 1825 as a public testimony to San Pablo's role in distributing cinchona first in Peru and then around the world, and it appears in many of the maps of that time.

The final decade of the *botica* at San Pablo saw Brother Salumbrino's ambitions come to fruition. The pharmacy itself, where the cinchona bark was weighed out and packed, was beautifully furnished. On its wall hung a large portrait of Salumbrino which his fellow Jesuits had commissioned in 1764 at a cost of 140 *pesos*, and which bore the legend: 'Agustino Salumbrino, first founder of this pharmacy of San Pablo'.

The walls were covered from floor to ceiling with solid oak shelves laden with bottles and flasks. Several tables and chairs were spread around the room, made of wood imported from Chile, and in the centre of the room was a long, wide mahogany counter of a beautiful reddish-brown colour. On top of the counter, in sharp contrast to the dark heavy wood, rested four delicate scales.

The three black employees who worked in the pharmacy spent their day in the laboratory, a forest of glazed earthenware and shiny instruments, some of lead or bronze, some of pure silver. The laboratory was filled with large jugs, scales, all kinds of stills used for distilling liquids, glass and metal funnels of all shapes and sizes, crystal flasks, retorts and matrasses, gridirons and hand mills, pumping engines and ovens, condensers and cauldrons, handsaws and sieves.

Brother Salumbrino's Jesuit masters might have been uncomfortable in that room, with its heavy fumes and thick, unpleasant odours of medicines and chemicals, but they would have been happy to know that in San Pablo's pharmacy he and his brother pharmacists had the means to preserve and restore the health of the hundreds of priests working in the field. The final inventory of the pharmacy includes more than five hundred medicines, in addition to the books in the library and the vast quantity of stills, bottles and other material that filled the laboratory's shelves. Of the medicines in the pharmacy, by far the most valuable was *una grande tinafa* – a great jar – of cinchona bark, which is valued at one hundred *pesos*.

Despite the excellence of its pharmacy, the small world of San Pablo was about to be engulfed in political events that were fuelled, as so often happens, by fear and greed. Secret orders had arrived from Madrid: the Society of Jesus was to be expelled from the whole of the Spanish Empire on the orders of King Charles III, who feared its swelling power and longed to own its properties and who finally, after many decades, had chosen to believe the Jesuits' enemies who had long tried to discredit them in the eyes of Charles and his court.

At four o'clock in the morning of 9 September 1767 the Viceroy, Don Manuel de Amat, had everything ready to carry out the King's instructions in Peru. Four hundred soldiers were stationed within the viceregal palace. In the dead of night a number of the most important men in Lima also arrived at the back door of the palace, summoned by a handwritten note from Amat that read, 'I need you for matters of great service to the King, and I warn you to come so secretly that not even those of your household would realise that you had gone out.'

Amat personally assigned the troops to go to the various Jesuit houses, and named the civil executioners of the royal decree. Knowing that San Pablo was the heart of all Jesuit institutions in Peru, the Viceroy chose the *oidor*, a judge in the Audencia of Lima, Don Domingo de Orrantia, and the *alcalde* – the Mayor – Don José Puente de Ibanez, to lead the soldiers who would occupy the college. Many of those men had studied under Jesuit masters in the cloisters of San Pablo. Many had attended mass in the college chapel, and had even served as altar boys there. Some had Jesuit confessors, and not a few were involved with San Pablo's many commercial enterprises.

Of the seven hundred men who left the Plaza Mayor to execute the royal wishes, more than three hundred followed Orrantia, walking in silence for the three blocks that separated the palace from San Pablo. By the time the royal troops with their weapons reached the college it was not yet five o'clock. Inside the Jesuit fathers lay sleeping.

Don Domingo de Orrantia himself had been a student at San Pablo. He knew the college and its customs well, and had even on occasion helped out in the pharmacy, packing up shipments of cinchona bark to be sent to foreign missions. He banged the large bronze knocker on the heavy wooden door next to the church entrance several times. When he heard steps approaching the door from the inside, he called in a loud voice for a confessor to administer the last rites of the Church to a dying person. It was a request he knew no Catholic priest would refuse.

The Jesuit doorkeeper swung open the door, only to be pushed aside by guns and bayonets. In a few minutes the college was overrun. Armed soldiers occupied the courtyards and stairways, and the frightened Jesuits were rounded up and herded into the chapel. Many thought they were about to die, and prayers came easily to their lips.

Don Domingo, standing near the altar, was deeply moved at the sight of his former teachers, and could not bring himself to read the royal decree. He asked the notary Franciso Luque to read it to the fathers instead. But Luque, a former San Pablo schoolboy, broke down and had to be excused. Orrantia had no choice but to read the King's despotic order to his friends and teachers himself.

San Pablo, a centre of learning for two centuries, served during September and October 1767 as a temporary prison for nearly two hundred Jesuits. Among them were Salumbrino's heirs, the four priests who ran the *botica*. On 27 October they were transported to Callao, where they were made to board the ship *El Peruano*. As the vessel drew slowly away from the pier, every man's eyes clung to the receding coastline.

The authorities were anxious not to lose the *botica*'s services, for it supplied all the Viceroy's subjects. Its fate was at first entrusted to another religious group, the Fathers of the Oratory of St Philip Neri, or the Oratorians as they were more commonly known. But no one could recapture Agustino Salumbrino's entrepreneurial spirit. The pharmacy of San Pablo fell into decline, and within three years it had disappeared altogether.

Its influence, however, did not fade so easily. By the end of the eighteenth century, nearly three hundred ships were arriving in Spanish ports from the Americas every year. About eighty of these came from Peru, not one of which failed to carry a consignment, large or small, of cinchona bark. Official imports of cinchona – those declared to the customs officials at the Spanish port of Cadiz – amounted to an annual total that was valued at more than ten million *reales*. To put this in context, the value of *cascarilla* imports in the 1780s was three times that of exotic hardwoods from the Indies, and amounted to nearly 2 per cent of all imports from South America, including gold and silver bullion. By that time the Jesuit initiative in discovering a treatment for malaria, and sending it halfway round the globe to the world from which they had travelled, had taken on a life of its own. Quinine no longer needed the priests.

The Quarrel – England

*'The Peruvian bark, of which the Jesuites powder is
made, is an Excellent thing against all sorts of Agues.'*

WILLIAM SALMON, *Synopsis medicinae* (1671)

A thick grey fog, the remnants of winter everywhere in the
southern hemisphere, always hangs low in the mornings along
the northern Peruvian coast. On 24 August 1748 the sun
struggled to find a break in the clouds. Nevertheless, the new
frigate the *Conde* seemed to be gleaming in its berth at the
dockside in Callao, north of Lima.

All over Callao's port, barefoot porters and barrowboys
elbowed and shoved their way up the gangplanks, their faces
red with the effort of carrying cargo of a hundredweight or
more on their shoulders to be stowed in the hold of the ship.
The *Conde* had been built in the yard at Guayaquil, a little
further up the coast, and had been launched into the Pacific
just days before. One hundred and twenty-four merchants
had clubbed together to charter her for this, her maiden voyage. The master of the *Conde*, Don Juan Basilio de Molina,
was well known to them as an experienced sailor who ran a
tight ship and knew how to care for his vessel during the heavy
storms it would encounter on its seven-month journey
through the Mar del Sur, the unpredictable and vicious southern ocean around Cape Horn, on its way to the port of Cadiz

in Spain. When, a few weeks earlier, a small contingent of the merchants had travelled to Guayaquil to make their final inspection of the vessel before signing the charter contract, they approved of her thick new sails, her heavy draught and her ample hold with room for all the merchandise they wanted to send to markets in Spain and the rest of Europe.

Of the charterers, thirty were women. This was unusual, though less unusual than it would have been a hundred years earlier. By the middle of the eighteenth century Lima was a thriving city. Thousands of Spaniards lived there, and with them their wives and children, as well as other European immigrants and Indians who had moved down from the Andes. Many of them intermarried with the local people, as had Martin Garcia Orãs de Loyola, the great-nephew of St Ignatius, the founder of the Society of Jesus, when he wed the Inca princess Beatriz Clara Qoya at the end of the 1590s.

The busy streets and markets may have given the impression that life was easy in Lima, but young families often struggled, and many people died young. Most of the women who had chipped in to help charter the *Conde* were widows, with no choice but to take up the chair at the head of the family table when their husbands died. In order to raise and educate their children, they took over their husbands' affairs and continued to run their businesses as best they could.

Standing on the dockside that late August day were three of the women who had helped charter the *Conde*, Dona María Josepha de Albinar, Dona Pheliziana Barbara Garzan and her friend Elena Woodlock y Grant, an Englishwoman who practised in Lima as a doctor, and who some years earlier had left Scotland with her husband, Don Juan Grant de Rufimurcus, one of the many John Grants of Rothiemurcus whose descendants still live in the Highlands today.

The three women, especially the fabulously wealthy Dona Pheliziana, accounted for a large portion of the cargo that was being loaded on board. As the fog began slowly to lift, they stepped forward to observe their ship, lifting their jet-black skirts and letting go of the black mantillas they had held tightly across their cheeks against the wind. One by one they counted the porters carrying their loads: silver in doubloons as well as bars – nearly two and a half million pesos' worth of bullion from the mines at Potosí – gold *castellanos*, sacks of fine vicuna wool, cases of cocoa, vanilla, white sugar, rich chocolate paste, leather skins and, of course, *cascarilla*, the Peruvian bark that the Jesuits had introduced to Europe as the newest, and by far the best, treatment for the tertian ague or the Roman fever.

Nearly all of the ship's two thousand *libras* of Peruvian bark was being loaded aboard on behalf of the three women. There would be heavy import duties to pay when it reached Spain – 10 per cent to the King, 1 per cent to the port of Cadiz, half a per cent to the *prestamistas*, the clearing agents, not to mention the quarter per cent offering that it was suggested should be made to *la Santa Iglesia*, the Holy Mother Church, but which everyone knew was a tithe that could not be refused. Yet, as a result of the success of the first small package of dried bark that Brother Salumbrino had sent to Rome in 1631, demand for cinchona was growing all over Europe. Dona Pheliziana and her friends knew that, despite the taxes, they would still make a handsome enough profit from the shipment to make all the risks worthwhile. Peruvian bark, which they bought for just under one peso a pound from the Jesuit pharmacy at San Pablo, or from other suppliers if they could find it more cheaply, could command twenty or thirty times that sum in Cadiz, and even more once it reached Rome, Amsterdam or London.

If the bark owed its early popularity to the Jesuits at the College of San Pablo in Lima, it was another Jesuit, across the sea in Europe, who had been its biggest champion.

Cardinal Juan de Lugo was born in Madrid in 1583, and after spending his early childhood in Seville went at the age of sixteen to the University of Salamanca to study law and jurisprudence. While continuing his studies, he became a year later a novice in the Society of Jesus, following the example of his older brother. At the age of twenty he wrote a legal treatise entitled *De Justitia et Jure*, which dealt with the practical problems of administering justice, and which demonstrated that he had a mind of his own and the courage to live according to his convictions rather than what his conservative seniors might dictate. De Lugo was a determined litigator who was not averse to contradicting precedent and pointing out error, and he expounded his conclusions with remarkable lucidity. The treatise, which became a landmark in the history of seventeenth-century law, might have opened many doors, and made its young author a wealthy man. But de Lugo was determined to become a Jesuit priest; and in any case, he knew that the many Jesuit centres of learning in Spain offered him the best environment in which to broaden his studies in philosophy and theology.

De Lugo's reputation spread, and when he was thirty-seven years old the Vicar-General of the Society of Jesus decided that he should go to Rome to teach at the Gregorian University. In 1643 de Lugo reached the age of sixty. His health, which had never been terribly robust, had been undermined during his years in the Holy City by the inevitable Roman fevers. And he felt that his austere and uneventful life was coming to its end.

But around that time *De Justitia et Jure*, which had been written forty years earlier, was republished. It was dedicated to Pope Urban VIII, who was impressed by the legal brilliance of the author's mind and his clarity of thought. Europe was exhausted by religious strife. The Thirty Years' War was raging, and Pope Urban's diplomats were engaged in a dozen or more other ecclesiastical disputes that had resulted from the Reformation. Men like de Lugo, with clear, visionary minds, were greatly needed in the Vatican.

The Pope summoned de Lugo, and informed him that he was making him a cardinal. Most of the priests who jostled for influence in the *curia* under Pope Urban and his predecessor, Pope Gregory, who had given his nephew Cardinal Ludovico Ludovisi so much power, would have jumped at such a chance. But the ageing Jesuit knew that his vows precluded him from accepting external honours. He also knew that he could not disobey his pontiff. De Lugo resolved his dilemma by bowing his head and obeying. But he was determined that his new eminence should not change his nature, and he made his investiture the first and last ceremony he ever attended at the papal court. After it was over, de Lugo returned to his former austere way of life. He used only a few rooms in the palace that had been given him in Rome, and reduced the number of servants in his house to the minimum. Most of his cardinal's income he gave away to charities.

Whether or not Cardinal de Lugo would have fulfilled the Pope's intentions for him will never be known, for seven months later, in the midsummer of 1644, Pope Urban VIII was dead. Like his predecessor Gregory XV, Urban died in

July. Once again, it seemed, a papal conclave would have to be held in this most malarious of European cities during the very worst season of the year. The physicians attending the cardinals urged that the conclave be postponed until the autumn. They received the unexpected support of King Louis XIV of France, for at that moment Cardinal Guido Bentivoglio, the man many thought would become the next Pope, was far away from Rome as the Papal Nuncio in Paris. However, Pope Urban's two Cardinal nephews were eager that their influence should prevail, and they persuaded the rest of the College of Cardinals to call the conclave at once. Bentivoglio would have no choice but to travel to Rome if he wanted to take part.

The threat of the summer marsh fever was uppermost in the minds of the cardinals who contemplated making the journey. It seemed to many that the dreadful conclave of 1623 might be about to repeat itself. In the event, three weeks after he arrived, Cardinal Bentivoglio succumbed to the Roman fevers.

The new Pope, Innocent X, a former Papal Nuncio in Madrid, was especially close to the Spanish cardinals, and had long been an ally of their King. Not that Cardinal de Lugo was in a position to take much notice of this at the time. Along with a number of the other cardinals who had attended the conclave he too had fallen ill with the fever, and it was many months before he felt well enough to take up his work once more.

In 1649, five years after the election of Pope Innocent, the Vicar-General of the Society of Jesus died, and the provincial fathers from all over the world congregated in Rome to elect a new head. As a cardinal, Juan de Lugo was one of the most important men present. The congress opened on 13 December, and among the delegates who had come from as far afield

as Poland, Mexico, Goa and Japan was Father Bartholomé
Tafur, the Peruvian Jesuit from San Pablo who had travelled
to Rome carrying a quantity of quinine bark in his luggage
and who, four years later, would become the head of the
college in Lima. Cardinal de Lugo, whose brother was now
with the Jesuits in Mexico, was eager to hear what St Ignatius'
followers were doing in the New World. He and Tafur shared
a deep interest in theology. Moreover, the two men had
already met before, when Tafur had come to Rome for the
eighth general congregation of the Jesuits which began in
November 1645.

What the Spanish Cardinal needed most, though, given the
recurring bouts of malaria from which he suffered, were the
fresh supplies Tafur brought of cinchona bark. Among his
many responsibilities, de Lugo had been appointed as the
director of the apothecary at the Santo Spirito hospital, whose
doctors had treated Pope Urban VIII after the terrible conclave of 1623. According to a Jesuit and naturalist, Honoratus
Fabri, de Lugo had the Peruvian bark tested by a Portuguese
doctor, Gabriele Fonseca, who was the new Pope's personal
physician. Unfortunately, no record of Fonseca's trials survives. But de Lugo was eager to begin distributing the bark
to the poor of the city. Within a short time, Pietro Paolo
Puccerini, a lay brother in the Jesuit order who had become the
director of the pharmacy at the Collegium Romanum, printed
a recipe in his *Schedula Romana* extolling the virtues of
cinchona bark as a remedy for tertian and quartan fever, and
explaining how it should be used:

> This bark comes from the Realm of Peru, and is called
> China or real China of the fevers, which is taken for the
> quartan and tertian fevers that are accompanied by cold. It

90

should be taken thus: two drachms, finely ground and sifted, three hours before the fever is due to take hold, should be made in an infusion in a glass of white wine. When the shivering starts, or when the slightest cold is felt, all of the preparation should be taken at once and the patient put to bed . . . For four days no other medication whatsoever must be taken. It must be used only on the advice of the physician who may consider whether it is timely to administer it.

The drug removed the sickness 'infallibly', Puccerini wrote, and although a few patients suffered relapses, these were readily treated with additional doses. Puccerini treated many thousands of patients each year, and so far as he was aware, he wrote, the bark had had no bad effects in Rome, Florence or the papal states. Moreover, he wrote, the bark had been conveyed to Naples, Genoa, Milan, Piedmont, England, France, Flanders and Germany.

Despite Puccerini's enthusiasm, the physicians and pharmacists of the time actually knew very little about the Peruvian bark. They probably would have had great difficulty in distinguishing between cinchona, especially if it was in powder, and most of the other foul-tasting dried barks that were stored in jars on apothecaries' shelves across Europe. The dosages they advised were based more on guesswork than on any empirical evidence; it was inevitable, sooner or later, that the wonder drug would fail to cure a case of tertian fever.

In the autumn of 1652 the bark was administered, following the instructions of Puccerini's *Schedula Romana*, to a patient of some eminence, the Archduke Leopold William of Austria, Governor of Belgium and Burgundy, who had been diagnosed as having a double quartan fever. The Archduke's own physician, Jean-Jacques Chiflet, had not been greatly in favour of the experiment, yet he gave in under pressure from other prominent members of the court who had heard tell of the bark's curative properties in Rome. For a while the Archduke appeared to be cured, but a month later the fever returned.

Had the Archduke been prepared to take another dose of bark when the fever recurred, he might well have recovered. But he was not willing to, in part perhaps because his brother, the Holy Roman Emperor Ferdinand III, had sent him a letter disapproving of its use. Indeed, the Archduke was so outraged that the bark had not delivered him at once of his malady that he decided to show his royal displeasure by ordering his physician, in defiance of the Vatican, to write a book denouncing the new remedy as a fraud.

Exposure of the Febrifuge Powder from the American World by Joannes Jacobus Chifletus, to give his full name in Latin, appeared in 1653, and was widely discussed, especially in the Protestant world, which was naturally suspicious of any remedy promoted by the Jesuits. In his account, Chiflet said that the bark merely lengthened the intervals between fevers but did not cure it; the drug was not needed in Europe, he added, for many other febrifuges were available. Moreover, he insisted it was dangerous, causing the humours to putrefy in the gall bladder, spleen, stomach and intestines, and to liquefy and spill over into adjacent organs, producing 'prolonged and fatal diseases'.

A prompt reply to Chiflet was published in Rome in the form of a book called *Pulvis Peruvianus Vindicatus de Ventilatore Ejusdemaque Suspecta Defensio*, or *Vindication of the Peruvian Powder*, in which the author claimed that in that same year, 1653, thousands of people in Rome had been cured of the fevers by this remedy. The book was signed with the name Antimus Conygius, a pseudonym for Honoratus Fabri, the same man who asserted that Cardinal de Lugo had had the Peruvian bark tested when it first arrived in Rome by the Pope's personal physician.

Cleverly defending the Jesuit order and its achievements, Fabri set out to destroy Chiflet's arguments. The authorities Chiflet cited had lived at a time when the bark had not yet reached Rome, he said, and were therefore irrelevant. Chiflet's interpretation of his experience was false, because it insisted that the powder had helped no one, and his reasoning was quite unsound. If the drug was unnecessary in Europe, as Chiflet maintained, there would be no need for many of the other febrifuges, such as myrrh and gentian, that he mentioned without disapproval. If Fabri so wished, he concluded, he could mention by name more than a hundred people who had benefited from the bark, including cardinals, princes, high government officials and 'persons of religion'.

Fabri's rebuttal did nothing to allay the wrath of Archduke Leopold, who hurried to find himself a more eminent voice than Chiflet's with which to shout down the Jesuits. He chose a Protestant-born physician and medical academic, Vopiscus Fortunatus Plempius, who had converted to Catholicism at the age of thirty-two, not out of conviction but in order to take up a prestigious job as Professor of Medicine at the University of Louvain, where he now enjoyed the wonderful title of 'Rector Magnificus'. Plempius, though officially a Catholic, was a Protestant at heart; he was also by nature an arch-conservative.

Some years earlier, Professor Plempius had taken pleasure in vilifying the English physician William Harvey when he published his discovery of the circulation of the blood, wondering loudly how Harvey dared to contradict the ancients. In taking on this fresh controversy, Plempius maintained that he had not understood a word of Fabri's arguments, but that did not bother him because he did not have the least desire to do so. He had read, he wrote, the writings of Galen, Hippocrates and Avicenna, who knew how to write about medicine and who never spoke of powders, let alone powders from Peru, as remedies. He quoted three cases where the drug had failed, and made the astonishing assertion that the bark had the effect of turning sporadic attacks into daily fever. Moreover, he added, in a vicious personal attack, he refused to believe that Cardinal de Lugo was supporting the use of the new remedy, since this eminent man was known to him as a great theologian, but in no other capacity could he be described as a fount of knowledge, medical or otherwise.

Religious bigotry played a large part in the arguments that raged about the bark's efficacy. Cinchona had been discovered by the Jesuits, and was being openly promoted by one of their most senior members – so much so that it was being called 'Cardinal's powder' or 'Jesuit powder'. That did not please the Protestants. But there was another aspect to the arguments. The raging controversy symbolised the growing divide between the hardline medical conservatives, traditionalists who wanted to preserve the ancient doctrine of Galenic medicine at any cost, and the younger, more adventurous doctors who were slowly coming to realise that Galen, master though he may have been, was wrong about many things.

Cinchona, more than any other drug, would be crucial in overturning Galen's teachings, which had survived for nearly two thousand years. It would change the direction of European medical thought.

Galen, a Greek physician who began practising medicine in Rome in 162 AD, incorporated the earlier ideas of Hippocrates and the ancient theory that all matter is composed of four elements: earth, air, fire and water. Diseases were thought to be caused by an imbalance of the four humours: blood, phlegm, black bile and yellow bile. Fever was a bile-caused disorder, and was considered a disease in its own right rather than a symptom. A patient with a high fever was said to be suffering from a fermentation of the blood resulting from too much bile. When fermented, blood behaved a little like boiling milk, producing a thick, frothy residue that had to be got rid of before the patient could recover. For this reason the preferred treatments for fever were bleeding or purging, or both. Tertian ague, which produced both the shivers and a very high fever, with nothing else to show for it – no vomiting or diarrhoea – was an extreme example of the blood fermenting. Yet patients (and their doctors) were supposed to believe that Peruvian bark, which was said to cure the fever, did so without producing any residue. How, a Galenist would ask, could this be so?

The study of natural history grew increasingly fashionable in England in the second half of the seventeenth century, and the gentlemen of the court vied to impress one another with what they knew. One session of the Royal Society in 1683 examined, in turn, the composition of chyle, widely considered to be the life-blood from which all human beings were made, the Arabic alphabet, the anatomy of the rattlesnake, the smell and colour of wine, and conducted a dissection of the genitals of a wild boar ('The penis was three-quarters of a yard long, crooked towards the end, winding about like a wimble,' wrote John Evelyn, the diarist, who attended the meeting). Despite an enthusiasm for the novel, the medical profession in England in the middle and late seventeenth century was deeply conservative, and still strictly Galenist.

Religious and social turmoil abounded during the reign of

Charles I and the nine-year rule of Oliver Cromwell which followed the King's execution in 1649. There was little scientific advancement. People were frightened, and English Protestants were reluctant to import anything that smacked of Catholic knowledge. In 1702, more than four decades before Dona Pheliziana Barbara Garzan and her sister merchants were sending Peruvian bark in commercial quantities from South America to Europe, a large supply of the most effective variety of bark then known, *Cinchona calisaya*, unexpectedly found its way to London with the capture of the Spanish fleet at Vigo, in the Bay of Biscay. In addition to nearly five thousand pounds of silver, the British buccaneers who overran the fifteen battleships also made off with thousands of pounds of pepper, cochineal, cocoa, snuff, indigo, hides and Peruvian bark. Before that, the only way for feverish English patients to obtain the bark had been from travellers from the continent who made their way across the Channel with small supplies of it in their luggage, probably Roman Catholics who had bought it either in Rome or in Belgium, by then the unofficial northern European centre of the Jesuit order.

These travellers were careful about selling the bark too openly because of its Catholic associations, yet the English, almost more than anyone in Europe, were in terrible need of the Peruvian remedy. If malaria was common around the Mediterranean, it was also prevalent in southern Britain in the seventeenth century. Kent, Essex, Somerset near Bridgwater, and parts of London, especially Lambeth and Westminster, were notorious for endemic malaria. '*London* Men of Pleasure' went to Essex, wrote Daniel Defoe in the early eighteenth century, 'for the Pleasure of Shooting; and often came home very well loaden with Game; and sometimes too with an *Essex* Ague on their Backs, which they find a heavier Load than the

Fowls they have shot'. Marriage to an Essex farmer was risky. Census figures show that at any one time there were nearly 12 per cent more adult males than females in the marsh parishes of Kent and Essex, and remarriage was common. 'All along this County,' wrote Defoe, 'it is very frequent to meet with Men that have had from Five to Six, to Fourteen or Fifteen wives; and I was informed, that in the Marshes, over-against *Canvy Island*, was a Farmer, who was then living with the five-and-twentieth; and that his Son, who was but Thirty-five Years old had already had about Fourteen ... The reason, as a merry Fellow told me, who said he had had about a Dozen, was this, That they being bred in the Marshes themselves, and *seasoned* to the Place, did pretty well; but that they generally chose to leave their own Lasses to their Neighbours out of the Marshes, and went into the Uplands for a Wife: That when they took the young Women out of the wholesome fresh Air, they were clear and healthy; but when they came into the Marshes among the Fogs and Damps, they presently changed Complexion, got an Ague or two, and seldom held it above half a Year; or a Year at most. And then, said he, we go to the Uplands again, and fetch another.'

Not that London itself was much healthier. The bill of mortality for the London area for the year 1665 included 5257 deaths from 'fever and ague', in a population of about 460,000. This was the year of the Great Plague, and deaths from that disease might have been deliberately concealed under a less fearful name. In any case, there is no need to search inaccessible records. Familiar books of the period, including Samuel Pepys and John Evelyn's diaries and Jonathan Swift's *The Journal to Stella,* as well as earlier works such as *The Canterbury Tales* and Shakespeare's plays, give plenty of information on the subject. Pepys' business was several times hindered by 'my lord's ague', and one of his merry parties was interrupted by the ague-fit of a lady guest. Earlier in the century Sir Walter Raleigh, during his captivity in the Tower of London, adjacent to the Thames, had 'prayed fervently to God that he might not be seized with an ague-fit on the scaffold, lest his enemies should proclaim that he had met his death, shivering with fear'.

John Evelyn has a good deal to say about Charles II's ague: he tells of one dangerous attack when the King insisted on having the new remedy 'quinquina'. He also describes his own little son's death from 'a quartan ague', and gives a full account of the popular cure – bathing the legs in milk as hot as could be borne and drinking 'carduus posset', a hot beverage made of thistle which, it was believed, would reduce fever, and with the help of which he cut short one of his own attacks of ague.

The most famous Briton to fall ill with a tertian ague in the 1650s was Oliver Cromwell. The Great Defender was born in the fens, the wet lowlands of East Anglia whose inhabitants shuttered themselves against the freezing northern winds from the North Sea in winter and suffered wave after wave of malaria in summer, when pools of stagnant water lay all around. He had spent much time in Huntingdon, in present-day Cambridgeshire, which was then recognised as being particularly malarious.

Although Cromwell is known to have suffered several times from the tertian ague, malaria, there is no evidence that he died of it – his death was most probably due to kidney stones and an infection of the urinary tract – nor, as tradition relates, that he refused to take the Peruvian bark on the grounds that he did not want to be 'Jesuited to death'. However, the fact that this myth is still widely believed to be true bears testimony to the strong feeling that abounded about the poisonous-tasting bark and its fervent Catholic supporters.

In 1658, nevertheless, the year of Cromwell's death, Londoners began to see discreet advertisements for the bark. On 24 June, for example, the *Mercurius Politicus*, the official publication Cromwell had approved three years earlier, included a notice which read:

> These are to give notice, that the excellent powder known by the name of the Jesuits Powder, which cureth all manner of Agues, Quotidian, Tertian or Quartan, brought over by James Thompson Merchant of Antwerp, is to be had at the Black-Spred-Eagle over against Black and White Court in the Old Baily, or at the shop of Mr John Crook, at the sign of the ship in St Pauls Churchyard, a bookseller, with directions for using the same.

Notices published later that year added that the bark was approved by 'Doctor Prujean, and other eminent Doctors and Physitians who have made experience of it'. Little is known of James Thompson, other than that he was an associate of Edward Somerset, Earl of Worcester, a leading Roman Catholic who was imprisoned in the Tower of London in 1652, but Sir Francis Prujean was the President of the Royal College of Physicians.

Whether Prujean really was a promoter of the Jesuit powder, or had given permission for his name to be so blatantly used, is not known. But another name was soon to come to prominence that would seal the reputation of the Peruvian bark, though not before it had greatly angered Prujean and his fellow physicians.

Robert Talbor was born in 1642 to a respectable family near Cambridge, close to where Cromwell was also born. His father worked for the Bishop of Ely, and his grandfather, James Talbor, had been the Registrar of Cambridge University. When he was still in his teens, Robert Talbor was apprenticed to an apothecary in Cambridge named Dent. One of young Talbor's tasks in Dent's apothecary was to study how better to administer the Peruvian bark as a cure for fever. At that time not much was known about how it should be prescribed, and many patients suffered from its side-effects, which can include tinnitus and, in pregnant women, nausea, headaches and vomiting.

Not long after, Talbor moved to Essex, where intermittent fevers were especially prevalent in summer. It may have been there that he met Elizabeth Aylet, the young woman he would later marry, for she came from Rivenhall, not far from Colchester. Like many in the seventeenth century, the Talbors spelled their name in several different ways. They had one son, who became an officer in the army and was known as 'Handsome Tabor'.

While Talbor *père* was living in Essex he perfected his method of curing the fevers, and in 1670 he set up shop in London as a *pyretiatro*, a fever specialist. Talbor had no medical credentials, so he wrote a book. He had the title set in Greek letters to impress his readers with his learning, and he opened with a witty poem:

> The Learned Author in a generous Fit
> T'oblige his Country hath of Agues Writ:
> Physicians now shall be reproacht no more,
> Nor Essex shake with Agues as before
> Since certain health salutes her sickly shoar.

Talbor was careful in *Pyretologia, or A Rational Account of the Cause and Cure of Agues* to disabuse anyone of the notion that he was a promoter of the Catholics' Peruvian bark. He cleverly disparaged it, saying: 'Beware of all palliative Cures and especially of that known by the name of Jesuit's Powder . . . for I have seen most dangerous effects follow the taking of that medicine.' In this he appeared to be conforming to the prevailing medical opinion, though he then contradicted himself by adding: 'Yet is this Powder not all together to be condemned; for it is a noble and safe medicine, if rightly prepared and corrected, and administered by a skilful hand . . .'

In its tone, Talbor's *Pyretologia* lies somewhere between a salesman's pitch and an objectively argued scientific report. He did not want to be seen to be giving credence to a Roman Catholic remedy, yet he also wanted to promote his own cure, which he insisted on keeping a secret from his fellow physicians and the public. It consisted, he said, of two ingredients from England and two from abroad. He refused to disclose any more, though he claimed it worked like a charm.

While Talbor hid behind this apparent medical conservatism, the cause of the Peruvian bark received a fillip in 1676 when no less a personage than Thomas Sydenham, known as the Hippocrates of English medicine, dwelt on the subject of the new drug in his *Observationes medicae*. Sydenham had begun to practise medicine in Westminster in about 1655, and before long he was one of the most influential physicians in London. He was obsessed by 'the fevers'. What were they? What caused them? In some cases they were accompanied by other symptoms, such as respiratory or intestinal difficulties. But then there was what was known as a 'pure' fever, where a patient who had previously been well suddenly began to shiver and shudder, then grew hot with fever, but had no other symptoms. What was that?

Sydenham was an important thinker, but he was also

cautious. Initially, he was sceptical about the Peruvian bark, but in the second half of the 1650s his friend and colleague Robert Brady, Regius Professor of Physic at Cambridge, began to use it. Gradually, Sydenham came to have more confidence in the treatment; later that same year he wrote to Brady: 'The Peruvian bark has become my sheet-anchor,' and insisted that it was the sole remedy for the quartan fever.

Supported by Sydenham, Talbor's practice was soon very fashionable. He charged fabulous fees, and refused to divulge the secret of his remedy. One can just see him, smiling and boasting, happy to have put one over the Royal College of Physicians, whose members continued to regard him as a quack – which of course is what he was. Yet, as Sir William Osler, the famed doctor and Professor of Medicine at Oxford University in the early twentieth century once observed, 'The trouble with quacks is that they cure people.'

In 1678 Talbor was summoned to treat the King. Charles II was a cynical soul. He had lived by then through the worst and best of times, and he knew that those around him would sell their mothers if they thought it would help enhance their influence at court. Years later, John Evelyn recounted how Richard Lower, the King's physician, had tried to have Talbor forbidden from administering his secret remedy to the monarch. But King Charles sent for Dr Short, to have 'his opinion of it privately, he being reputed a papist . . . he sent him word, it was the onely thing could save his life, and then the King injoyn'd his physitians to give it to him, and was recovered'. Ever happy to retail gossip, Evelyn went on to add that when Short was asked by the King why the other doctors would not prescribe it, 'Dr Lower said it would spoile their practise or some such expression'.

The Royal College of Physicians was even more distressed

when, not long after, it received a letter from Lord Arlington, the King's Secretary of State, which read: 'His Majesty, having received great satisfaction in the abilities . . . of Dr Talbor for the cure of agues has caused him to be admitted and sworn one of his physicians; and, being graciously inclined to give him all favour and assistance . . . for the public good, has commanded me to signify his pleasure unto you and the rest of the college of physicians that you should not give him any molestation or disturbance in his practice.' That was only the beginning of the College's discomfort and of Talbor's extraordinary success: on 27 July 1678, King Charles knighted him at Whitehall.

Soon after, Talbor left London for France. Whether he decided to abandon the city where the alleged 'Popish Plot' had just been revealed and anti-Catholic sentiment was growing steadily, or whether he was despatched by the King, who heard that the Dauphin, the only surviving son of Louis XIV, was dying of fever, is not known. Nonetheless, Talbor was soon busy in Paris curing the French notables of the fever that plagued their city nearly as much as it did London. When the Queen of Spain fell ill, he was sent by Louis XIV to the Spanish court. Having cured her, he returned to Paris to much acclaim; his secret remedy became the great vogue of the day, and he was addressed as 'Chevalier'.

Five years before Talbor reached Paris, Molière had staged *Le Malade Imaginaire*, which attacked Antoine Daquin, the King's private physician, and the rest of the French medical establishment. He ridiculed the mannerisms of the physicians – their wigs, their fancy clothes, their speech – but also the mummified Galenic ideas to which they clung. And the people laughed. The court laughed. They laughed most of all at the final mocking scene, in which the doctor is conferred with his degree: 'Dost thou swear to follow the advice of the ancients whether good or bad? Dost thou swear not to use any other remedy but those used by the Faculty [of Medicine of Paris] even if the patient should die?' With his unorthodox medical ideas – his secret remedy and his refusal to purge or draw blood – 'Chevalier' Talbor was the living negation of that medicine that Molière, in his genius, mocked so successfully. 'The remedy of the Englishman makes them quite contemptible with all their bloodletting. It is to take life itself away from them to take fevers from their domain,' wrote Madame de Sévigné, commenting on the discomfort of the physicians at Talbor's success. 'If only Molière were alive.'

Having witnessed how successful Talbor was at curing the sick, the French King offered to buy his secret remedy. Talbor agreed, but only on condition that the King maintain the mystery until after Talbor's death. The doctor was awarded three thousand gold crowns and a substantial pension for the rest of his life. Two years later, he died at the age of thirty-nine.

In January 1682 Louis XIV had a book published revealing Talbor's secret. *The English Remedy; or, Talbor's Wonderful Secret for Cureing of Agues and Feavers* was immediately translated into English, and became a bestseller. Talbor's formula, it turned out, was made up of six drachms of rose leaves, two ounces of lemon juice and a strong infusion of Peruvian bark, all of it administered in wine. Talbor was careful always to change the wines he used, in order to disguise the identity of his cure. His real secret was to administer a weaker dose than patients had hitherto been accustomed to, and to repeat it often if necessary.

Two years before Talbor died, another advertisement, this time in the *London Gazette*, stated that 'the infallible Medicine of Sir Robert Talbor, for cure of Agues and Feavours, being in his absence rightly prepared by his Brother . . . is to be had when he is out of Town, at Mr Lords a Barber in St Swithings-lane . . . The price a guinea two doses.' That Talbor could charge such high prices in London, where once he had been laughed off as a quack by the medical establishment, was a sign of the extent to which the Peruvian bark had become an established and successful remedy in England.

Shortly before he died, Talbor erected for his family a monument in the churchyard at Trinity College, Cambridge, on which he described himself as the 'most honourable Robert Talbor, Knight and Singular Physician, unique in curing Fevers of which he delivered Charles II King of England, Louis XIV King of France, the Most Serene Dauphin, Princes, many a Duke and a large number of lesser personages'.

His enemies still referred to him as 'a debauched apothecary's apprentice', a 'seller of secrets' and an 'ignorant empiric'. Whether they disliked the ease with which he moved in Roman Catholic circles, or whether they just disliked him, the fact remains that it was Talbor, more than anyone else, who single-handedly ensured that the Jesuit bark would never be forgotten in England. Many were now prepared to try it on any fever, especially if no previous medicament had worked. Nowhere was this more evident than in the last days of King Charles II's life when, with their monarch racked by fever and in great pain, the royal physicians were willing to try almost anything.

The King had awoken on Monday, 26 January 1685 pale and feeling shaky. His barber came into his room to shave him, but no sooner had he attached a towel to one side of the monarch's face and was passing behind his chair to fix the other than the King collapsed with a seizure. Help was at hand, though. It was the custom that no one, on pain of death, was permitted to bleed the King without the consent of his senior ministers. But his doctor, Sir Edmund King, defied tradition and insisted he be bled at once. Sixteen ounces of blood were removed from a vein in the King's right arm. He remained in the barber's chair, his teeth held forcibly open to prevent him biting his tongue, while his numerous other physicians were called. They ordered cupping glasses to be applied to his shoulders before proceeding to scarify him, by which they removed another eight ounces of blood.

A strong emetic of antimony was then administered, but the King could be persuaded to swallow only a little of it, so they gave him a full dose of sulphate of zinc as well as some strong purgatives and a succession of clysters, an early kind of suppository or enema. His hair was shorn, and pungent blistering agents were attached all over his head. Amazingly, given such treatment, Charles soon recovered consciousness, and his doctors pronounced him out of danger for the moment.

In the evening, the King's doctors met once more to discuss how to relieve the 'humours' on his brain. They gave him remedies to make him sneeze, as well as a preparation of cowslip flowers and sal ammoniac, or smelling salts, and applied noxious plasters to his feet.

Early the next morning another consultation was held, with no fewer than twelve physicians present. The King complained of a pain in his throat, and his jugular vein was drained of a further ten ounces of blood. By the following morning the doctors were so pleased with the progress of their treatment that they contented themselves with ordering the patient to take a therapeutic rest for the day, which at the time was a most unusual prescription.

Despite this, the convulsions returned in the evening, and

Charles was given a drink known as 'Spirit of Human Skull', a fashionable concoction the ingredients of which remain uncertain. The physicians noticed that there was a cycle to the convulsions, and some of them suggested that the King was suffering from an intermittent or tertian fever. The remedy, they insisted, was the Peruvian bark.

The Privy Council, which was by now quite confused by the many and varied diagnoses that had been suggested, were informed. One of the Councillors, Lord North, asked, 'Could anything be worse?'

One of the physicians answered him: 'We now know what to do.'

'And what is that?' asked North.

'To give the cortex,' replied the doctor, meaning the Jesuit bark. The attending physicians did as he suggested, but to no avail. Within two days the King was dead.

That quinine had failed to save Charles II's life did not, as might have been expected, destroy the reputation of the miracle drug. Talbor's success with the Jesuit powder, and the blessing that the English and French kings had given to this exotic cure, had already done much to spread its renown and ensure that it would not quickly fall from fashion. But as quinine grew more popular, new problems developed. The greater the demand, the more likely it was that unscrupulous merchants would adulterate pure quinine with other bitter-tasting barks, or sell foul substitutes for it. The most vexing problem, then, was how to ensure that adequate supplies of pure quinine reached Europe.

Talbor may have succeeded in immortalising the bark, but it would be other men who would cross the ocean to seek out the tree from which it came, which no one in the Old World had yet seen.

The Quest – South America

*'The Peruvian bark should be taken with a very little
quantity of Laudanum; Children should take it in
Chocolate well sweetned.'*

SIR HANS SLOANE, President of the
Royal Society, 1727–41

Six months to the day after the death of Charles II, Sir John
Evelyn paid a visit to the Chelsea Physic Garden.

In the twelve years since it had been laid out in the village
green on the north bank of the Thames, the garden had become such an attraction for Londoners that its owners, the
Worshipful Company of Apothecaries, had been forced to
build a brick wall around it to keep out interlopers. In 1680,
John Watts was appointed to manage the garden. Watts was
no ordinary gardener. Trained as an apothecary, he was also a
plantsman, an entrepreneur and a prosperous merchant who
in time would forge the Chelsea Physic Garden's links with
horticulturists in a number of other countries. But the first
task he set himself was to build a heated greenhouse. With
this he hoped to expand the garden's production of herbs such
as mint, sage, pennyroyal, sweet marjoram and rue, and
'foreign as well as native plants' including 'nectarines of all
sortes, Peaches, Apricotes, cherryes and plumes of several
sorts of the best to be Gott'.

Early greenhouses, so named because they were meant to conserve evergreens in winter, were stone or brick buildings with tiled roofs and glass windows all along one side. They had long been in use throughout northern Europe, where the winters were long and only the hardiest of plants survived outside. But heated greenhouses were something quite new, and Mr Watts' new example was what Evelyn wanted to see most. His drinking companion Sir Hans Sloane, the naturalist and later generous patron of the Physic Garden, who once summoned his friends to help him dissect an elephant on the lawn in front of his Chelsea house, had been to see it a year earlier, and had been waxing lyrical about it ever since. In a letter to a friend Sloane wrote that Watts 'has a new contrivance. He thinks to make, by this means, an artificial spring, summer and winter.' Later Sloane reported to the same friend that the greenhouse had proved highly successful, and that the severe winter had 'killed scarse any of his fine plants'.

On 6 August 1685 Evelyn wrote in his diary: 'I went to Lond: to see Mr Wats, keeper of the Apothecaries Garden of simples at Chelsey, & what was very ingenious the subterranean heate, conveyed by a stove under the Conservatory, which ws all vaulted with brick; so as he leaves the doores & windowes open in the hard[e]st frosts, secluding onely the snow & c:' Equally, if not more, interesting to our eyes over three centuries later is what Evelyn found in Watts' new greenhouse: 'a collection of innumerable rarities [. . .] Particularly, besides many rare annuals the Tree bearing the Jesuits bark, which had don such cures in quartans'.

Evelyn was understandably intrigued. Not more than a hundred books on herbs and plants had been published in Europe at that time, and hardly any of them contained any details of the flora to be found in the furthest corners of the

Spanish Empire. No one could travel to South America without the express permission of the King of Spain, and nothing on South America could be published without it first being submitted for censorship to the Holy Office of the Inquisition, the Council of the Indies and the Casa de Contratación, which controlled all trade with the Spanish world. Every bookseller in Spain was required to furnish a list of all the books he offered for sale, and was forced to destroy any that the Inquisition condemned; such was its power. As a result, Europeans knew as little about South America, its people, its landscape and its natural life as they knew about the moon.

Evelyn's throwaway observation of 'the Tree bearing the Jesuits bark' is the first mention we have of a live cinchona plant outside South America. It is not to be found in the famous *Gardeners Dictionary* written by Watts' successor Philip Miller and published in 1731; botanical historians assume that by then the cinchona tree in the Chelsea Physic Garden was dead.

Nor is anything known about how it arrived at the Chelsea Physic Garden, or who brought it. Transporting plants across long distances in the seventeenth century was far more difficult than transporting dried bark and keeping it from becoming mildewed. The journey, by mule and then by sea, was long and arduous, and the cinchona, which thrived in the damp, forested climate of the high Andes, would not have taken easily to the hot, salty humidity of life at sea level. The glass-domed Wardian case that would later be used to keep plants healthy onboard ship without wasting precious supplies of fresh water had yet to be invented. But someone had taken the trouble to care for Mr Watts' precious tree, watering it and keeping its roots and leaves in good order throughout a journey that would have lasted the better part of a year. Or perhaps the explanation was far simpler: gardeners and horticulturists often collect seeds while travelling abroad, and the enterprising European plantsmen of the late seventeenth century were no exception. Perhaps the tree was germinated in Chelsea from seeds brought home in someone's pocket.

The effort was a pointer, though, to things to come. For the story of quinine over the next two centuries is overwhelmingly about the naturalists who crossed the ocean to find the cinchona tree and study it in its natural habitat; and, later, about the merchants who used all their ingenuity and guile to acquire its seeds, even stealing them and smuggling them abroad on occasion, in order to plant cinchona elsewhere and harvest it on a commercial scale.

Peru, Ecuador and Bolivia – where the cinchona tree grew in the wild – were all part of the Spanish Empire, yet it was not a Spaniard who first went in search of the cinchona tree, but a Frenchman. And he began by wanting to measure the circumference of the earth.

Charles-Marie de la Condamine was a wealthy and well-connected Parisian, with thin lips and flared nostrils that would have given his face an air of haughty disdain had it not been for his open smile and his unflagging enthusiasm for everything that was new. He had become well known at an early age for his passion for science and mathematics, but his particular fascination was for geodesy, the geometric measuring of the earth's surface.

Shortly after la Condamine was elected to France's Académie Royale des Sciences, that august body became embroiled in a dispute over whether the earth was flattened at the poles or elongated, and more egg-shaped. The argument had begun with Sir Isaac Newton, who believed the earth was a globe that grew flatter the further north you travelled, and bulged at its middle. Ranged against Newton and his followers was the French Astronomer Royal, Jean Dominique Cassini, who long before had evolved a theory which held that 'Man infests a globe which lengthens in the direction of the polar diameter. The world is a prolate spheroid, lengthened at the poles, pulled in at the equator, much as a pot-bellied man might pull in his girth by taking in a few notches in his belt.' Like la Condamine, Cassini was a member of the Académie Royale des Sciences. Newton was trespassing on the Académie's territory. What was more, he was an Englishman.

But the issue was far more than one of intellectual machismo or simple national pride. At stake was the future of navigation, cartography and trade, even of science itself. Cassini and the French navy, together with the Académie, had spent much time trying to ensure that their charts were as accurate as possible. To achieve this they needed to know the exact length of a degree of latitude. Through triangulation, another academician, Jean Picard, in 1669 measured the meridian of an arc of one degree between Sourdon, near Amiens, and Malvoisine, south of Paris, and found it to be 69.1 miles. But this measure could be universally applied only if the earth were a perfect sphere. It was clear that further proof of the planet's shape was needed.

Picard completed his measurements at the same time that Christian Huygens, a Dutchman, brought his newly patented pendulum clock to Paris. In 1672 Cassini sent Picard to French Guyana, on the north-east coast of South America, with one of Huygens's clocks, only for Picard to discover that it beat more slowly there. In order to keep perfect time at the equator, the pendulum of a clock set in Paris, at 49° latitude, had to be shortened. The reason, as we now know, is that the earth's bulge at the equator decreases the pull of gravity. In an effort to settle the matter once and for all, the Académie determined to measure the length of a degree of latitude at the North Pole and at the equator, and compare the results.

They had no trouble in deciding on Lapland to stand in for the North Pole, but where should they choose for the equatorial measure? Africa was still largely unexplored, the lower Amazon was a quagmire, Borneo unopened. Much of the rest of equatorial Asia lay within the influence of the Dutch, mercantile rivals and Protestants to boot. Solidly Catholic Peru was the answer, and Charles-Marie de la Condamine the man to lead the expedition, not least because he was prepared to pledge 100,000 *livres* of his own money to help finance it. From then on, when his friend Voltaire wrote to him he would address him always as '*Mon très ambulant philosophe . . .*'.

In order to travel through South America, la Condamine had first to secure the permission of King Philip V of Spain. Despite great opposition from the Council of the Indies, who feared – rightly as it turned out – that a French expedition to the interior of South America might in the long run undermine Madrid's power there, the King acceded. But he insisted the expedition be accompanied by two Spanish naval officers, Captain Jorge Juan y Santacilla, a mathematician, and the King's spokesman, the twenty-year-old Captain Antonio de Ulloa, another noted mathematician, who would one day be appointed Governor of Louisiana. Their job was to report back to Madrid on the Frenchmen's activities as well as on the 'state of the colonial empire'. Secret instructions were sent to all the places they might visit: local officials were to give the French visitors all the help they needed, but they were to be careful not to permit them to '*poner los ojos en la Tierra*' –

in other words, they would not be allowed to see too much.

As it turned out, the Spaniards found nothing untoward to report. But the King had also wanted, Ulloa wrote, 'to give his own subjects a taste for the . . . sciences'. In that, Philip V had been inspired. The expedition would foster in Spain a longing for intellectual as well as physical expansiveness, for geography, biology and botany, and a yearning for the country to take its place among the great scientific nations of the world. That yearning, in turn, would greatly influence the study of the cinchona tree in years to come.

In July 1735, after a nine-week voyage from La Rochelle, la Condamine disembarked with his instruments at Cartagena de las Indias, in what is now Colombia. Cartagena was already a great fortress city, and along with Puertobelo in Panama and Vera Cruz in Mexico, one of the three gateways to the Americas. In addition to the two Spanish officers, la Condamine was accompanied by ten other Frenchmen, including a young botanist named Joseph de Jussieu, who began right away to scour the surrounding hills for fruits and flowers with which to begin a botanical collection. His role in acquainting the world with the cinchona tree would be as tragic and ill-fated as la Condamine's would be fruitful and lucky.

In March of the following year, having reached Guayaquil on Peru's north-western coast, the same port where Dona Pheliziana and her fellow charterers would commission the construction of the *Conde* twelve years later, la Condamine unpacked his precious measuring instruments for the first time. It was frustrating work. The heavy, drizzling *garúa* fog was almost permanently settled on the coast, which meant that the sun was visible only in the evening, and not in the morning, depriving the measurers of 'the correspondent observations of which we were in want'. The cloudy sky and constant rain prevented them from observing the eclipses of the satellites of Jupiter, and only a brief break in the weather permitted them to observe the end of a complete eclipse of the moon. In addition, the food was poor and the insects were a constant menace. Soon the expedition members began to find each other very irritating.

La Condamine was more than happy to take a break from measuring and accompany a new friend, a mathematician and cartographer named Don Pedro Maldonado, up the Río Esmeraldas to the interior of the Audencia de Quito, the region that lay to the east and north-east of the viceroyalty of Peru. Surrounded by exotic birds like the toucan with its ceaseless song, '*Dios te dé, Dios te dé*', and vegetation such as he had never seen, la Condamine quickly became fascinated by the region's plants and trees. An Indian brought him a strange piece of stretching 'cloth' called *caoutchouc*. This was rubber. He watched the Indians gather milk from scars they had cut in the local hevea trees, and was amazed at how the latex solidified. Noticing that the new material was water-resistant, he fashioned a covering out of it for his Hadley octant and other instruments to keep them safe from the humid tropical air. Later he would be the first man to carry samples of rubber back to Europe.

His head full of new discoveries, la Condamine put his fellow measurers out of his mind. Accompanied by the adventurous Maldonado, he penetrated deeper into the jungle that flanked the Río Esmeraldas. The explorers' instruments lay at the bottom of their forty-foot dugout, in between bananas, long, parsnip-shaped *yuca*, rice, beans and several chickens, their legs tied to prevent them escaping, which from time to time would rise up and cluck in loud vexation. The travellers drank *masato*, a pale sugary beer made from fermented plantain mash. At Maldonado's insistence, la Condamine traded in his tattered stockings and grey greatcoat for a pair of local woollen pants and a poncho, learning to gather it about him when he climbed upwards to avoid tripping over it. La Condamine mapped the courses of the Esmeraldas, collected plants for the expedition botanist Joseph de Jussieu, and picked up a shiny metal which was neither gold nor silver, but was known by the local name of *platino*.

Soon the explorers left the river valley and began climbing. Local guides, small men who painted themselves entirely red, led them through the Esmeraldean forest towards the hills. Rain fell without cease, usurping the sun's place as the measure of time. Above their heads was green, underfoot it was black, with thick mucky sediment. At every step their feet sank into ooze. Little light penetrated the dense foliage; mud and fallen trees carpeted the trail, except where a running brook turned it into a quagmire. Drenched, miserable, hungry and exhausted, la Condamine's only solace was the protection offered to his instruments by the rubber cloth the Indians had given him by the Esmeraldas.

Slowly the party emerged above the jungle, and the plants around them began to change. Gone were the thick, matted forest and the tightly laced lianas. La Condamine had no need of a thermometer to tell him that the temperature was falling, and that he was entering a different isometric zone; the proof was there in the greenery surrounding him. On the steaming plains below, the explorers had seen bananas and palms. As they climbed, these gave way to tree ferns, red and purple gentians and thin trees with gnarled, twisted trunks covered in moss.

And then, on 14 February 1737, while exploring the area around Loxa armed with notes supplied by de Jussieu, la

Condamine saw for the first time the tree the locals called *quinquina*, from which the Jesuits derived their miraculous bark. He sat by the roadside and sketched a branch with its leaves, flowers and seeds. The drawing would accompany the memoir, *Sur l'Arbre du Quinquina*, he sent back to the Académie in Paris, which was published in its journal in 1738. It was the first description Europeans had had of the living tree, and when the great Swedish taxonomist Linnaeus read of la Condamine's observations, he gave the *quinquina* tree a new name, *Cinchona*, as a tribute to the legendary Countess of Chinchón.

La Condamine writes of three types of cinchona, which he distinguishes as white, yellow and red. He was told by his local guides that the trees' barks had differing febrifuge qualities, with the white possessing scarcely any, and the red being the most potent. Other than that, there was no botanical difference between them; they all grew together in the same locality, although never in a mass, but always isolated from one another and always surrounded by trees of other species. When allowed to attain their prime, they grew taller than most of the trees surrounding them, with a trunk that could attain the thickness of a man's body. Already, little more than a century after the Jesuits had begun shipping the bark to Europe, la Condamine noted that the large trees from which the first bark had been stripped were dead, and that nearly all the trees around Loxa had been destroyed except the youngest.

La Condamine's botanical descriptions were vague, and his drawings lacked precision. But although the exact identity of the species he described would be debated for years afterwards, he is remembered as a hero. The botanist Joseph de Jussieu, however, whose detailed notes first led la Condamine to find the miraculous fever-tree, is now all but forgotten. In 1761 his vast collection of plant specimens, laboriously gathered over a quarter of a century, was stolen in Buenos Aires by a servant who thought the boxes contained money and other valuables. For the next ten years nothing is known of de Jussieu's movements, but by the time he returned to Paris in 1771 he was insane, unable to remember even his own name. In 1936, to commemorate the centenary of its foundation, a French company that manufactured quinine under the name '*Trois Cachets*' published a manuscript by de Jussieu. With one minor exception, it is the only work of his ever to have been published. It is a highly detailed description of virtually all the different major species of *Cinchona*, and was written in Latin in 1737, the same year la Condamine saw the fever tree for the first time.

In 1774, the French once again petitioned the King of Spain to send a botanist to South America. Louis XVI's Chief Minister, Anne Robert Jacques Turgot, had two objectives in mind. The first was to try to recoup the material that had been assembled by de Jussieu, whose ill luck was still being lamented by his family and friends at the French court, and the other was to add to the collections of plants that had been pouring into France over the previous decade from India, the South Seas, Madagascar and the Cape of Good Hope. There was, however, almost nothing from South America, and the gap would not easily be filled without Spain's permission.

The Spanish King, Charles III, had learned a great deal from the experience of la Condamine's expedition, and the idea of a major voyage to South America captained by a non-Spaniard left him far from happy. The Englishman Captain James Cook was just completing his second circumnavigation of the globe, and Spain's scientifically-minded monarch had ambitions to see his kingdom at the forefront of the scientific enlightenment that was flourishing in Europe. Earlier that same year, the King had ordered that a new botanical garden be established on the Prado in the centre of Madrid, to rival similar gardens that were being opened in Vienna, Versailles, Frankfurt, Budapest and Coimbra in Portugal.

He had also just bought one of the finest natural history museums assembled in Europe, the cabinet of curiosities of Pedro Francisco Davila, a highly educated man who came from Guayaquil but who had lived for nearly a quarter of a century in Paris. When the museum opened to the public in Madrid in 1776, with Davila as its first director, it boasted two magnificent rooms for minerals, two salons of stuffed animals and birds, another hall for displaying insects, and yet another for sea life. A big room was set aside for 'exquisite rare woods', and in the middle of it was a stuffed elephant and its skeleton.

Despite these treasures, the museum was not yet full, and word went out to all the Spanish realms that more exhibits were needed. In particular, the museum required specimens of trees: cinnamon, Paraguayan tea, *jalapa* (a Mexican climber that yields a purgative drug), cedars, ebony, white balsam, black balsam and the rare tropical hardwoods like *campeche*, *cocobola*, *violeta*, *moradillo*, *paloferro*, *granadillo* – many of whose names appear on the shipping inventory of the *Conde*. Last, but not least, the museum wanted specimens of the precious cinchona tree from which came the Jesuit bark, which was now known to grow in a sweep across the northern part of the South American continent, from the Pacific to the Atlantic. What better way could there be for Spain to show off its intellectual prowess than to promote a generation of travellers who would bring South America's scientific trophies back to Europe on its behalf?

Seen from afar, the exploration of South America might have appeared a relatively simple matter. But as with the first expedition, this new venture would become fraught with tension and troubles – troubles that grew ever more intractable the deeper it penetrated into the interior of the continent.

The party that set off from Cadiz on 4 November 1777 sailed in one of the best ships in the Spanish fleet, the sixty-cannon *Peruano*, the same vessel that had carried the Peruvian Jesuits away from Lima a decade earlier. They took with them a substantial library, including several works by Linnaeus and the compilations of some of the earlier explorers in America.

The aim of the expedition, as ordered by Charles III, was 'the methodical examination and identification of the products of nature of my American dominions, not only to promote the progress of the physical sciences, but also to banish doubts and falsifications which exist in medicine, painting and other important arts, and to foster commerce, and to form herbaria and collections of the products of nature, describing and making drawings of the plants found in these, my fertile dominions, in order to enrich my Museum of Natural History and the Botanical Garden of the Court'.

The expedition was led by Don Hipólito Ruiz Lopez, a twenty-three-year-old pharmacist and botanist who had worked for a short time in the King's botanical garden. Ruiz was a small man with a cupid's mouth and high arching eyebrows. His hair was brushed down in waves over his forehead, giving his face an air of anxiety. He was already suffering from lung disease and from a depression which only grew worse as he worried about the task that lay before him. For, as one of his sons would later write, Ruiz was 'very zealous for the glory of the nation'.

With him was another trained pharmacist, José Antonio Pavón, and Joseph Dombey, a French botanist appointed by Louis XVI. Dombey would grumble a great deal at being paid less than his Spanish colleagues, though they were less inclined to sympathise with him once they discovered that he described them patronisingly in his letters as 'my two Spanish pupils'.

Completing the party were two painters, whose task it would be to make reproductions of the flora. They had been selected from among a number of art students in Madrid for their 'bachelorhood, skill and a gentle disposition', and were commanded to 'copy nature exactly, without presuming to correct or embellish it, as some draftsmen are accustomed to do by adding colouring and adornment taken right out of their imagination'. Despite every effort, the publication of the *Flora Peruviana et Chilensis* would prove a nightmare, and there were many on the expedition who would wish they had never left home.

Arriving in Lima after an ocean crossing that lasted nearly six months, during which only one other ship was sighted, the expedition found a city of a little over fifty thousand people, dense with monasteries and markets selling all manner of tropical fruits, from papayas to pepinos, and five different kinds of banana. But what most astounded the Europeans was the grinding poverty of the Limenos' everyday lives, and their longing for little items of luxury to help overcome their difficulties. 'One sees many ladies riding in their carriages,' Ruiz wrote in his journal, 'the mules of which have more to eat than the family. Many wear diamonds but have no daily bread, except when they visit the pawnbrokers. Great numbers of women are loaded down with fine trappings, yet their children go bare. Countless people are dazzlingly dressed, although they are in debt for every stitch in their bodies.'

For nearly two years the foreigners foraged around Lima, making drawings and collecting specimens. In April 1780, when the summer torrents in the *montaña* began to die down, they set off, trailed by a string of pack mules, for the mountains, and the botanists' virgin land.

The journey started gently enough, but it became more difficult when they turned north-east into the rocky valley beyond La Oroya. They fought to make themselves heard against the roaring cascade of the Rimac, the same river that made its way to the sea beyond Lima, but fell silent after they saw one of their mules trip and fall into the icy water that immediately swept it away. All but one of the local drivers deserted on the upward climb, and the scientists were forced to become muleteers. Ruiz was fascinated by the Indians he met along the road, and spent much time studying the various methods they used to dye their bright woollen clothes.

After travelling more than three hundred miles, the party finally reached Huánuco, a small town that lies on the forested eastern slope of the Andes overlooking the jungle. Ruiz was little impressed by Huánuco, which had once flourished as the centre of the feudal estates around. The frogs that lived in the irrigation ditches chirped and croaked all night, and the guinea pigs were host to 'a tremendous plague of minute fleas, unbearable because of their bites'. Nor was he much taken by the Indians who burned dried manure to keep their huts warm. 'This is why the houses are always full of smoke and stink unbearably,' he wrote in his journal in May 1780.

Ruiz spent much time thinking about how the lives of the Huánuco Indians might be improved: by importing a strict bishop to care for their spiritual life, and by increasing the cultivation of cocoa, coffee, indigo, balsam, vanilla and other tropical products. He pinned most of his hopes, though, on the *quina*, the quinine-producing bark which grew in the forests surrounding Huánuco, and which rivalled in quality the bark that had so far mostly been found around Loxa, to the north, in what is now southern Ecuador. The three botanists quickly forgot their troubles when they entered the *quina* forests for the first time. Here, more than three years after they had begun their journey, was the most important tree they had been charged to find.

They started their exploration of the cinchona forest by setting up camp at Cuchero, a small village set on 'a small, level shelf on a mountain' about fifty miles north-east of Huánuco. Early next morning they pressed on. Ruiz wrote in his journal that they 'crossed a high, narrow slope, dangerous because the pathway up was of sharp rock and very rugged. We spent the night under a boulder, where the three of us could hardly fit. We sat all through the night without a wink of sleep because of the ceaseless roar of the rushing brook and the incessant croaking of the infinite number of little frogs that breed in its waters.'

The next day, he wrote, 'we travelled about a league and a half through a trail full of brambles, creepers and other weeds growing across the path from one side to the other. We had to pick our way carefully to avoid falling and getting hurt or being cut to pieces by the spines and prickles of the plants. As a result of the rainstorm the night before, the road had become impassable. The mules could not walk without slipping continuously, though some were so skilful that they put their two front feet together and purposely allowed themselves to slide down on sloping ground. Adding to the difficulties of the road itself were the tangles of branches hindering progress, the narrowness of the hillside paths, and the constant climbing uphill and down.' Yet Ruiz would not give up. 'Half a league onwards,' he wrote, 'we entered thick woods, on a trail full of ruts and holes . . . In this locality I discovered *Cinchona purpurea*, [the] purple quinine tree, the first that I had examined up to that time.'

The nascent *quina* industry that Ruiz observed around Cuchero had been started just four years previously, when a passing trader named Don Francisco Renquifo noticed cinchona trees growing wild in the forest. He had previously seen similar trees around Loxa. 'He gathered some samples of bark,' Ruiz wrote, 'took them to Huánuco and showed them to a number of people. He told them that the inhabitants of Loxa carried on a considerable business in this bark, and he insinuated that they, too, might make large sums of money from the stands of *cascarilla* trees he had found.' In the year before Ruiz's visit the area had produced seventy-five thousand pounds of bark, which was sent to a dealer in Lima.

The harvest was, however, causing serious depredation to the forest, as collectors stripped branches and cut down trees with little heed for the future. The advice that had been carefully passed down by the Jesuits, of planting five saplings for every tree that was cut down, was being ignored in the rush to cash in on the demand for still more bark. The visitors also realised, as they walked through the forest, that much *quina* was being left behind. 'We botanists who travelled about in these forests witnessed the considerable waste that the bark gatherers left by not stripping all the bark from the stems and trunks because these parts did not have commercial value,' wrote Ruiz.

It was thus that the Spanish botanist, following de Jussieu's

Harvesting cinchona, as it would have looked to Charles-Marie de la Condamine and Joseph de Jussieu when they visited the Peruvian Andes in the late 1730s.

Scion of a French botanical dynasty, Joseph de Jussieu *(far right)* would go insane in his bid to be the first European to see the cinchona tree in the wild.

José Celestino Mutis, Spanish physician and botanist, whose death in 1808 ended any hope of making sense of the magnificent *Herbarium* he had amassed at the Botanical Institute of New Granada.

Linnaeus, the eighteenth-century botanist, who misspelt the name
of the Countess of Chinchón and erroneously named the tree *cinchona*
in his system of plant taxonomy.

ZORN. IC. PL. MED.

Cinchona officinalis,
Cinchona calisaya and
Cinchona ovata as drawn by
eighteenth- and nineteenth-
century travellers.

Cinchona officinalis. L.

William Balfour Baikie, the Orkney doctor whose unpublished manuscript on the wonders of quinine describes his success at quelling malaria while exploring the River Niger in 1854, thus opening up west and central Africa to exploration and eventual colonisation.

My great-grandfather, Philippe Bunau-Varilla, who overcame both yellow fever and malaria in the 1880s during the unsuccessful French attempt to build the Panama Canal.

example, hit upon the idea of making quinine extract. He advised the producers in Huánuco to macerate a quantity of chopped-up bark in four parts of water for about forty hours, then to cook it over a low flame until half the liquor had been consumed. The remaining liquid was to be filtered and then cooked once again, until it attained the consistency of caramel-like resin, before being stored in boxes made of the quinine trees themselves, carefully sealed to keep out the humidity. 'Since that time,' Ruiz would proudly write twelve years later, 'others followed their example and more than forty thousand pounds of extract had been shipped to Europe and, as word of its efficacy and lower price spread about, the business stood to increase even more.'

Although Dombey would soon return to France, Ruiz and Pavón stayed on in South America for another eight years, travelling as far south as Concepción in Chile, where they identified and named *Araucaria araucana*, the Chilean pine or monkey-puzzle tree, with its deep green spines and its strange configuration of branches. Their explorations were highly successful, but their attempts at making complete notes of what they saw suffered terrible ill luck, as did their efforts to enlarge the King of Spain's botanical collections.

A first shipment of specimens, which had been put together even before they left for Huánuco and the *quina* forest around Cuchero, was lost when the ship in which it was being carried, the *Buen Consejo*, was captured by a British galleon and its contents dispersed. Then, in 1785, five years after the botanists saw the *quina* tree for the first time, by which time the two artists on the expedition had completed a significant number of drawings, all their manuscripts were lost when the huts in which the party lived and worked in the jungle at Macora, near Huánuco, were destroyed in a fire. Faced with an 'absolute volcano', Ruiz was desperate. 'I recklessly entered the fire where I knew my papers to be,' he wrote to a friend, 'but all was in vain. Circumstances forced me to get out, for the fire was still alive. Going half-crazy, I wanted to kill myself. But finally, overcome by rushing around and shouting, I fell on the ground at midnight exhausted.' The only thing that was saved from the fire was one of Ruiz's three parrots, which managed to escape with just its feathers singed.

Two years later, after the botanists had carefully packed and labelled a new consignment of specimens for shipment to Spain, another jinxed vessel, the *San Pedro de Alcántara*, ran into trouble. Even before starting its Atlantic crossing, the ship lost more than thirty containers of living plants, swept overboard in the stormy waters off southern Chile. Then, after the captain had carefully repaired his ship in Rio de Janeiro and replaced the deserting mariners, the *San Pedro de Alcántara* went onto the rocks off the jutting coastline that abuts Peniche, Portugal. Nearly two-thirds of the passengers and crew were drowned, and all of Ruiz and Pavón's remaining plants and specimens were lost.

The two botanists returned to Spain in 1788. Despite the losses, they still managed to bring back from Peru and Chile three thousand plant descriptions and more than two thousand drawings, enough for a dozen botanical volumes. The King who had commissioned the *Flora Peruviana et Chilensis*, Charles III, had died shortly before their return, but his son Charles IV was very keen that the task his father had left him should be completed. The book must breathe 'grandeur and magnificence' worthy of His Majesty, the young King insisted.

But the project continued at a snail's pace. Painters and illustrators who were capable of conveying grandiosity and magnificence were difficult to find, and worked very slowly. Money was also a problem. The Spanish exchequer had suffered during the recent war with the English, and there was little left over for extravagances, even if they were ordered by the King. A special fund-raising exercise was organised throughout the Indies, with contributions being sent from as far afield as Cuba, Venezuela and the Philippines as well as Chile and Peru. Yet even this was marred by delays and disputes, and the continuing disagreement between Joseph Dombey and the Spanish botanists would swell into a bitter quarrel over which nation, France or Spain, was the true owner of the drawings and specimens on which the plates of the *Flora Peruviana* were to be based. At one point this even threatened to put an end to the enterprise altogether. By 1791, only about a dozen plates of the *Flora* had been completed. In order to placate the authorities, Ruiz set about publishing a small volume on the precious cinchona tree the following year.

Quinología, o tratado del árbol de la quina ó cascarilla, con su descripción y la de otras especies de quinos nuevamente descubiertas en el Perú, to give its full name, was the first treatise on the miraculous cinchona tree to be published in Europe since la Condamine's description had been read out to the Académie des Sciences in 1738. Ruiz dedicated *Quinología* to Count Floridablanca, the Secretary of State who oversaw the construction of the King's botanical garden in Madrid, who 'experimenting more than once in the alleviation of [his] precious health through the efficacious beneficence of this Spanish specific [had] contributed by this means to increase more and more its esteem among men'.

No purveyor of medicines ever made such flamboyant claims for a tonic as did Ruiz in support of the miracle remedy *quina*. The bark could be ground up and infused or boiled, and administered as a liquid. Reduced to powdered form or as an extract, it could be given in pills, in preserves, in diluted wine or in water. Ruiz had seen it work, he said, against 'simple or complex intermittent fevers, malignant putrefactions, nervous malignancies, exanthemas and [the] putrid effects of small-pox'. He would even prescribe it to help cure, among other things, toothache, gangrene, dysentery, measles, miscarriages, headaches and collapsed lungs. Indeed, the only ailments against which it didn't seem to work, according to Ruiz, were gout and rheumatism. Given how much trouble he had gone to in order to obtain specimens of the tree, its leaves, flowers and seeds, it is hardly surprising that Ruiz wanted it to be a miraculous remedy that would cure almost any disease known to man. He, like everyone at that time, knew nothing about what malaria really was, how it was contracted, and how exactly quinine worked in treating it; nor would such things be known for at least another hundred years.

However far-fetched his claims for the miracle tree from across the ocean, Ruiz was level-headed enough in describing the many varieties of *quina* that he had seen in exact detail. He wanted, more than anything, to lay down the foundation for better botanical knowledge by announcing that he had observed, collected, described and had drawings made of seven species of cinchona, including *Cinchona officinalis*, a magnificent forty-foot tree which he believed to be identical to those from Loxa, and *Cinchona purpurea*, which though it had purple leaves, looked very similar to the *officinalis*. *Purpurea*, which grew most happily on the lower slopes of the Andes, was more abundant than *officinalis*, but contained almost no active quinine. That did not prevent unscrupulous collectors from trying to deceive dealers by mixing it with the *officinalis* bark, which was much rarer and more difficult to harvest.

Ruiz's slim *Quinología*, and the supplement to it that he completed nine years later with his colleague Pavón, would ignite a controversy over the taxonomy of cinchona, how the different species should be named, and which were the most productive, that would rage on for a hundred years. His rival was another Spanish botanist, José Celestino Mutis, publication of whose work would prove if anything to be even more fraught and tragic than that of Ruiz.

José Celestino Mutis studied medicine, but his only real ambition was to travel to the New World. Born in 1732 and raised in Cadiz, Spain's busiest port, he spent much of his youth hanging around the docks. Like Cardinal Juan de Lugo, whose interest in the New World had also been aroused at an early age in Cadiz, Mutis loved to scan the horizon and wait for the arrival of galleons like the *Conde*, or the ill-fated *San Pedro de Alcántara* that ran aground off the coast of Portugal with Hipólito Ruiz's precious cargo of specimens aboard. For hours on end he would watch the porters scurrying up and down the gangplank while others in the hold hoisted up cases of silver doubloons and gold *castellanos*, sacks of fine vicuna wool, planks of *campeche* and other rare tropical hardwoods, boxes of cocoa, vanilla, white sugar, rich chocolate paste, leather skins and, of course, the precious *quina* bark.

In 1757, when he was twenty-five and freshly qualified after years of studying medicine, Mutis travelled to Madrid to try to persuade the authorities to send him abroad. While he waited he passed his time collecting plants and grasses around the capital, for he was a born naturalist; only occasionally did he see patients, and then only to ensure he did not starve. Three years later, the King appointed a new Viceroy, Marqués de la Vega de Armijo, to the New Kingdom of Granada, now part of Colombia. Mutis asked for a position as a botanist, telling de la Vega that he wanted to make a systematic study of the forests of Jesuit bark. After giving the matter some thought, the Viceroy agreed, though he told Mutis that in order not to disturb the King over such a trivial matter, he should be attached to de la Vega's household as his personal physician.

Realising that this was the only way he might advance his project, Mutis agreed, and in February 1761 the new Viceroy made his official entrance into Santa Fe de Bogotá, the capital of the New Kingdom of Granada. Full of excitement about what he thought would be the imminent start of his grand botanical project, Mutis wrote a letter to the most eminent botanist in the world, the Swede Karl von Linne, or Linnaeus, as he is commonly known. That Linnaeus had never heard of him was entirely irrelevant to Mutis. Writing as fast as he was able, in a flowery Latin hand that grew more and more illegible with every page, he wanted only to tell the great man of his plan to write nothing less than a new natural history of the entire continent of South America, and to apprise him that he planned to make regular shipments of the rarest plants in the very near future.

Linnaeus answered Mutis immediately. He treated the young man as though he were already an illustrious colleague, gave him encouragement, requested his help and thanked him warmly for the forthcoming plants. He told him that he would work the tree, when the specimens arrived, into his great taxonomy under the name *Cinchona*, which he had given the genus after reading the description la Condamine gave the Académie des Sciences in 1738. (The misspelling of the word that should have been '*Chinchona*' arose from the fact that Mutis and Linnaeus corresponded in Latin.)

Lastly, Linnaeus sent Mutis's heart soaring with the news

that he had arranged to have the young Spaniard appointed a member of the Academy of Sciences of Uppsala. Thus began a correspondence that was to last until Linnaeus's death seventeen years later.

After such an encouraging start, however, Mutis would find that progress on the ground in Bogotá was harder to achieve. His friend the Viceroy, who had seemed so enthusiastic about the botanical project back in Spain, was taken up by other worries over his new job. Mutis worked hard as a physician, and soon found that he was one of the most popular doctors in the city. But the Viceroy, he realised, was ignoring him. 'While in Spain,' he admitted in anguish to his diary, 'I believed that I should already be on my way to Loxa to study the cinchona tree . . . so assuredly the Viceroy promised me that shortly after our arrival here he would send me on this errand.'

To distract himself, Mutis began teaching mathematics at the University of Bogotá. He took holy orders. He badgered the Viceroy's officials about when he might start his botanical studies, until in desperation in 1763 he undertook to write to the King himself. Charles III had not yet bought Pedro Francisco Davila's cabinet of curiosities, nor had he given the order to establish the botanical garden in Madrid. After a certain amount of flattery, Mutis pointed out to the King that Spain was neglecting the study of the natural sciences, in which other countries were making such extraordinary progress, and emphasised that Spain should have the best natural history museum in the world. He reminded Charles that his forebears had recognised the importance of the study of natural history, that Philip II and Ferdinand VI had sent scientists to America, but that the work they had started was far from done.

Mutis courteously but firmly outlined Spain's duty to the world in having been awarded by Providence such enormous possessions. 'America,' he wrote, 'is not only rich in gold, silver, precious stones and other treasures but also in natural products of the greatest value . . . There is quinine, a priceless possession of which Your Majesty is the only owner and which divine Providence has bestowed upon you for the good of mankind. It is indispensable to study the cinchona tree so that only the best kind will be sold to the public at the lowest price.' He warned that if the practice of chopping down trees to obtain the bark were to continue, the supply of the drug would be exhausted in a short time.

So convinced was Mutis that he was right, and so desperate was he to start work, that he ended by threatening the King with punishment from above, describing the spectacle of Charles III seated on his throne surrounded by the ghosts of those who had died for lack of the remedy.

The King did not reply.

A year later, in 1764, Mutis sent the King another petition, and again no answer came. He also sent Linnaeus some specimens from a cinchona tree that had been presented to him by the superintendent of the cinchona forest in Loxa. Linnaeus wrote back that he was most grateful for these samples, which allowed him to complete his classification of the tree whose name he had unwittingly misspelled. 'I was delighted,' he wrote, 'with the beautiful drawing of the bark and the leaves and flowers that you sent me, which flowers . . . gave me a clear impression of a very rare species.'

The man who gave Mutis the specimen from Loxa also told him that he had seen the tree growing in the New Kingdom of Granada. If this was true, it meant that the tree could survive north of the equator. Even more important, it might be as readily available on the Atlantic coast of South America as it was on the Pacific, which would greatly ease the problems of transportation.

All of this set Mutis on fire. Again he went to the Viceroy, who, while he acknowledged the importance of the botanical study of the tree, threw up his hands and declared he could do nothing, as the King had repeatedly ignored Mutis's petitions. De la Vega did however allow his desperate physician the occasional day exploring the outskirts of Bogotá. This was a limited concession indeed, for there was no cinchona to be had outside its natural habitat or close to the city.

As a distraction, Mutis threw himself into silver mining. In 1766, five years after he arrived in South America, he went to live in a shack by a dilapidated silver mine, where he addressed himself to the problems of improving labour conditions and extracting ores. For four years he remained there, until in 1770 he returned to the capital, though he had not quite given up his interest in mining. Thus it was that one day when he was travelling between Bogotá and the silver mine during the rainy season, he noticed a grove of cinchonas, with their unmistakable corona of pink and yellow flowers. The first person to whom Mutis sent examples of this species, which he named *Cinchona bogotensis*, was Linnaeus.

The discovery of cinchona trees growing so far north of the equator, and 150 miles east of their regular habitat, convinced Mutis that the tree would be found on the Atlantic side of the continent. Hearing that a new Viceroy was being appointed to the Kingdom, Mutis once again petitioned the authorities. De la Vega's successor was sufficiently convinced by the botanist-physician to write to Madrid urging the Crown to consider establishing a monopoly of the quinine trade, in the hope of eliminating smuggling and ensuring that the trade in the drug was efficiently organised, so that quinine from Peru was exported westwards towards the Philippines and the Orient, while that of New Granada was shipped eastwards towards Europe.

Yet again, the answer from Madrid was complete silence. In 1778 Mutis left Bogotá once again for a far-off silver mine. This was the same year that Ruiz and Pavón and their French colleague, Dombey, arrived in South America. If they had known of the difficulties Mutis had already suffered, they might never have made the journey.

Mutis was still working at the same silver mine five years later when another Viceroy, an archbishop by the name of Antonio Caballero y Gongora, paid a pastoral visit to the remote spot where he lived. Gongora was a scholar and a patron of science, and he was astonished to find Mutis in the wilderness of the New World. When Mutis told him the sad story of his failures in South America, the Viceroy's astonishment turned to fury. He took Mutis with him back to Bogotá, and immediately began to deluge the authorities in Madrid with petitions, reports and recommendations. Without waiting for a reply, he set about organising a provisional scientific commission entitled the Botanical Expedition of the New Kingdom of Granada.

Gongora's dogged persistence paid off. In November 1783, more than three years after Ruiz and Pavón had visited the quinine forests of Huánuco, Mutis was made by royal decree the chief botanist and astronomer of the new botanical expedition. By order of the Crown, his debts were paid off and he was awarded a pension of two thousand *pesos* a year to continue his studies. Orders were given that all the books and instruments necessary for the expedition were to be sent from England, where the best of them were made.

The first task that the expedition set itself was to send botanists to explore the forests of New Granada, in search of the cinchona tree. Headquarters were set up in a tiny town called Mariquita, and Mutis gave himself up to an orgy of botany. In no time the expedition began to accumulate manuscripts, specimens, and above all an endless number of hand-painted illustrations. The artists, who came from Quito, worked like medieval monks, for nine hours a day in complete silence. Their drawings were life-sized and elaborately detailed, showing all the phases of the plants' growth, including the seeds and roots. In order to obtain the most accurate representations, Mutis devised different ways of preparing dyes from natural sources. He also put in order all the materials he had collected from his various pilgrimages over the years; thus the expedition founded a unique library.

In 1791, at the same time that Ruiz after years of ill luck finally brought out his small treatise *Quinología*, the library and the collection of specimens of the Botanical Expedition of the New Kingdom of Granada moved from Mariquita to larger premises in the capital, Bogotá, and the expedition was renamed the Botanical Institute of New Granada. The staff now consisted of ten scientists and fourteen artists. Mutis made provision for any of the children in the orphanage of Bogotá who showed an aptitude for botanical drawing to be given proper training and the chance to join the institute at the end of their schooling.

Up to 1792, however, nothing had been published by the Botanical Institute. Material was constantly being accumulated, notes taken, specimens mounted and illustrations completed. Everyone was busy arranging the two major works that Mutis had planned: a monumental treatise on the flora of northern equatorial South America, which included thousands of illustrations; and a complete medical and botanical treatise on the cinchona tree, which was also to be profusely illustrated. Yet somehow neither work was ever quite ready for the printer.

In 1793 the Spanish government gave orders for an investigation of the Botanical Institute to be carried out. Having seen the difficulties that Ruiz and Pavón had experienced, the government was particularly anxious that Mutis's work should not be lost. The results of the investigation were far from encouraging: although the material gathered was clearly priceless, there was little prospect of publication, because not a single manuscript was anywhere near ready – there were only hundreds and hundreds of illustrations and thousands upon thousands of botanical descriptions, all scattered in confusion. If anyone should have been able to put some order into the work, it was Mutis, but the investigators were frankly unsure whether he was able or willing to complete the task.

To fend them off, Mutis undertook, much as Ruiz and Pavón had done a few months before, to publish two short pamphlets on the cinchona tree.

From the results, it is clear that Mutis had little interest in the pamphlets. Both deal largely with medical problems relating to the administration of cinchona, and he describes only four species of the tree, compared with Ruiz and Pavón's seven. The attention he pays to the drug's medical qualities is perfunctory to say the least, writing as he was decades after both Robert Talbor and Francesco Torti had published their clinical studies in exhaustive detail, and the pamphlets added little that was new.

José Celestino Mutis died in 1808, at the age of seventy-six, without publishing a page of the colossal mass of manuscripts and illustrations that he had accumulated throughout his life. Francisco José de Caldas, one of the most gifted of the boys from the orphanage in Bogotá to whom Mutis had given the opportunity of training and a salaried job, and with whom Mutis had worked closely on the cinchona study, was appointed a director of the institute. On orders from Madrid, he immediately set about arranging Mutis's work for publication.

Not many months had passed, however, before there was an insurrection in the New Kingdom of Granada. The rebels, under Simón Bolívar, wanted to wrest New Granada from the Spanish Crown, and although a provisional government was put in place, the royalist forces were determined to fight back. Caldas continued his work preparing Mutis's work for the printer. Before long, though, he was ordered to become a captain of the engineering corps, and to organise the building of cannon, gunpowder mills, ammunition factories and fortifications.

The insurrectionists fought the Spanish soldiers for five years. For a while it looked as if the rebellion might succeed, but the Spanish reorganised their forces, and finally entered Bogotá. Caldas was arrested and condemned to death. When the sentence was passed, he said nothing in his defence. He did, however, turn to the court and remind them that there was something more precious than his or any man's life: the work that had accumulated in the Botanical Institute, including the study of the cinchona tree on which the lives of so many people depended.

Mutis's work would be of no value to anyone, he emphasised, since he, Caldas, was the only person who could put it in order. Because of the years he had spent at Mutis's side working on the botanical treatises, he was in a position to publish Mutis's work. He asked the court that the time required to do this be granted to him before the death sentence was carried out. His request was denied. Refusing to give up, he sent a written petition to the authorities stating his case in further detail. Again the answer was no.

On the eve of his execution, Caldas was asked to give more details about Mutis's work, with a view to possible publication. Once again he asked for a postponement of the death sentence – only six months, and he would work with a chain tied to his ankle if necessary. A third time it was denied him.

The next day, 29 October 1816, Francisco José de Caldas was executed with his back turned on the firing squad, as traitors are often made to do.

Bolívar's forces were threatening Bogotá once again, and a year later, on orders from Madrid, all the material from the Botanical Institute was taken to Spain. Some was lost, but in all 5190 coloured illustrations and 711 drawings, forty-eight boxes of botanical specimens, mostly of the cinchona tree, and a large number of manuscripts, including Mutis's *Herbarium*, in which he had catalogued more than twenty thousand plants, were deposited in the cellars of the Jardín Botánico in Madrid. They were packed in 104 large wooden boxes, and while some of the work has been published in recent years, many of the catalogues that Mutis and Caldas prepared have never been opened.

To War and to Explore –
From Holland to West Africa

*'Our stores were nearly empty, and little or no Peruvian
bark – one of our most essential medicines – remained.'*

SIR JAMES MCGRIGOR, of the Army Medical Board,
reporting to the House of Commons investigation
of the Walcheren expedition, 1810

The death of José Celestino Mutis in 1808 represented the end
of the search for knowledge for knowledge's sake in the story of
the cinchona. Political rivalry, religious pressure, scientific
one-upmanship and petty human jealousy had all had a part to
play in the quest for the magical tree that produced the Jesuit
powder which cured the ague or intermittent fever. But the
principal inspiration, stretching over generations from Mutis
back to la Condamine, de Jussieu, Sydenham and Sloane, had
been the quest for knowledge. From the time of Mutis's death,
that driving force would change to a quest shaped by the
growth of commerce and political expansion.

In 1809, the year after Mutis died, Europeans would
witness at first hand what malaria could do to an army unpre-
pared. So many British troops died during the Walcheren
expeditionary raid on Napoleon's troops in Holland in the
summer of that year that the word 'Walcheren' is still listed in

some dictionaries as a byword for military blundering. The British Foreign Secretary, George Canning, strongly opposed the expedition, and challenged his cabinet colleague, the War Minister Lord Castlereagh, to a duel over it. When Canning lost, he was forced to resign.

Those most affected by Walcheren were the Royal Navy and the British Army. But the forces of other countries could not fail to heed the warning. From the start of the nineteenth century onwards, finding a cure that was cheap and easy to administer would exercise many who saw at first hand how malaria affected both warfare and exploration.

The Walcheren expedition took shape just as France was beginning to recover from the humiliation of its defeat at the Battle of Trafalgar in October 1805. By the spring of 1809, the French navy was well on the way to finishing the construction of its new fleet. As many as thirty-five freshly built ships lay at anchor as the spring sun returned to the Scheldt estuary on Holland's south-west coast, with only a few still on the stocks at Antwerp waiting to be completed. Napoleon, who kept up with the latest news of this swelling arsenal less than forty miles from the British coast, described it as 'a pistol held to England's head'.

In London it was realised that something had to be done, but no one was quite sure what. The cabinet was split, as were many of Britain's most senior generals. Orders were issued and then countermanded. Nevertheless, by the end of July, forty thousand soldiers were ready to be conveyed to Holland. It was the biggest expeditionary force that Britain had ever sent abroad. To transport them, an imposing fleet of thirty-eight ships of the line, thirty-six frigates and over six hundred other vessels – small ships of war, gunboats and transports – was assembled off the English Downs. When at last they sailed away in three divisions, to the sound of massed bands playing the national anthem, it seemed to one observer that an entire forest of masts, bearing a single canopy of white canvas, was on the move. 'Never will I forget the glorious sight,' an unknown Scottish soldier wrote in his diary, 'of the most powerful and numerous fleet which ever left the British shores. The sea looked as if it groaned under the weight of so many vessels, and as far as the eye could reach a wilderness of masts was seen.'

The slow and haphazard way in which the expedition had been organised through the early summer of 1809 meant that news of its impending departure had, inevitably, reached Napoleon's ears even before it set sail. The French Emperor's reaction was characteristically insouciant. He knew all about Walcheren and the fever that reappeared there every summer among the polluted polders, slimy, oozing ditches and dykes, and marshy dunes. 'Do not attempt to come to blows with the English,' Napoleon wrote to one of his commanders. 'Your National Guards, your conscripts, organised into provisional *demi-brigades*, huddled *pêle-mêle* into Antwerp with no officers for the most part and with an artillery half-formed . . . will infallibly be defeated. We must oppose the English with nothing but fever, which will soon devour them all. In a month the English will be obliged to take to their ships.'

Not even the most trusted of his commanders knew that when the time came, Napoleon himself would usher in the worst epidemic of malaria that Europe had ever seen by ordering that the dykes be breached to flood the whole of the Scheldt estuary with rank, briny water, the perfect environment for the mosquito larvae to hatch.

The expedition set off with a braggardly confidence, but little thought had been given to provisions and medical supplies. This was not unusual. Military medicine then was a ragbag affair. There was no Army Medical Corps. Physicians held no military rank, and most of those who followed the army combined their trade with a more lucrative private practice at home. Surgery was a grim, rough-and-ready business, and army surgeons had no more prestige than barbers. Medicines were often bad and dear, and apothecaries amounted to little more than energetic hawkers. There was no ether, no chloroform and certainly no modern antibiotics. Wounds to the limbs were dealt with mostly by amputation, and nearly two-thirds of surgical patients died either from loss of blood or from infection. Despite those terrible figures, by far the biggest killer at the end of the eighteenth century was not injury sustained in battle or on the operating table, but disease, particularly dysentery, typhoid and malaria.

The medical provisioning of the Walcheren expedition was shoddy even by the standards of the day. Neither the Physician-General, nor the Surgeon-General or the Inspector-General of Army Hospitals was consulted about how best to equip the fleet. Because the troops were travelling only a short distance, the duelling Lord Castlereagh informed the Military Secretary that there would be no need for hospital ships. For forty thousand men there were only thirty-three medical staff with any training, and just thirty untrained assistants to attend them.

Medical supplies too were virtually non-existent, which was all the more extraordinary considering that the British Army knew at first hand about the intermittent fevers that were endemic to the Zeeland coast. Nearly a century earlier, Sir John Pringle, an army physician whose public support of Captain Cook's campaign against scurvy had made him the father of military medicine, saw four-fifths of the army fall sick in the Lowlands campaign during the War of the Austrian Succession.

John Webb, a physician who accompanied the Walcheren expedition, described the unhealthy Zeeland environment in some detail:

> The Bottom of every Canal that has communication with the Sea is thickly covered with an Ooze, which when the Tide is out emits a most offensive and Noisome Effluvia; every Ditch is filled with Water, which is loaded with animal and vegetable Substances in a state of Putrefaction . . . The Endemic Diseases of the Country, remittent and Intermittent Fevers begin to appear about the middle of August . . . It is computed that nearly a third of the Inhabitants are attacked with Fever every sickly Season.

Webb's report notwithstanding, the troops sailed for Walcheren with barely a day's dose of Peruvian bark, one of the essential medicines for fighting the intermittent fever. Some ships carried no bark at all. One of those for whom there was no treatment was Webb himself. Within a few days of arriving he had succumbed to the fever, which hastened the end of his army medical career.

Among those who crossed the Channel to Holland was a young rifleman, John Harris, who had been much impressed by the send-off given the fleet from England, and who had his first glimpse of the sandy banks of the Dutch coast through the dawn mist on the morning of 29 July 1809.

Rifleman Harris was a Dorsetshire sheep-boy who in the long winter nights in the West Country had learned the art of making shoes. He had been drawn by ballot into the army, but lost his heart shortly afterwards to the dashing appearance of a detachment of the green-jacketed 95th Rifles. On catching sight of them – 'as reckless and devil-may-care a set of men as ever I beheld, before or since', he would write later – the young soldier immediately volunteered, carrying his tools and leather with him as the battalion's cobbler.

Harris served during the bombardment of Copenhagen in 1807 and in the terrible campaign in Spain, where the men found themselves with 'beards long and ragged; almost all were without shoes and stockings; many had their clothes and accoutrements in fragments, with their heads swathed in old rags, and our weapons covered in rust'.

A few months after returning to England at the beginning of 1809, Harris was called upon to go to war again. He was looking forward to a quieter time of it in Holland. As the expedition made its way up the shallow, winding forty-mile gap that separated Antwerp from the sea, the helmsmen manoeuvred with caution to avoid running aground upon the sandbanks of Walcheren island at the mouth of the estuary. Soon the smaller ships' boats were able to begin towing flat-bottomed craft towards the beaches. The soldiers leapt out into the shallow water, each man burdened with sixty rounds of ammunition, two spare flints, three days' rations of bread and cooked pork, an allowance of rum, a canteen and a rolled overcoat.

For a while Rifleman Harris thought he had found the quiet place he was looking for. One officer, dashing up the empty beach, saw outstretched before him a whole fertile island, 'a flat fen turned into a garden'. So taken aback were the officers and men alike who joined him that there was a chorus of 'How beautiful!' Another officer wrote home of the island's capital, Flushing: 'It is a very pretty little town and, like everything in this island, so clean and new that it gives you more the idea of the model of such a place, made of paste-work, than the town itself.' As Harris too later recalled, 'The Rifles occupied a very pretty village, with rows of trees on either side its principal streets, where we had plenty of leisure . . . The appearance of the country was extremely pleasant, and for a few days the men enjoyed themselves much.'

Appearances were misleading, though, and the mood soon changed when the weather broke. In the first week of August it began to rain. Gales blew in from the North Sea, rain fell hard and steadily as the men worked to build huts with wood and straw in which to shelter. Transporting the troops ashore had taken so long that there had been little time to follow up with the equipment they needed. Many were ill prepared against the weather, with no blankets or change of linen. Sporadic gunshots, fired during light skirmishes against the enemy, occasionally pierced the air. But for the most part the English troops were looking to take care of themselves.

On 10 August, on Napoleon's orders the French breached the sea dyke. By nightfall the water in the ditches where the English were building gun and mortar platforms had risen by three feet. The rain helped the flooding, and the sluices on the east side of the island had to be opened to let out as much water as possible. But at high water sticks were still needed to mark the firm ground. The whole island seemed to have been engulfed. In the ditches the men, wearing their greatcoats, were hot and soaked through. They worked round the clock, slapping about through the mud and water in shoes with rotten uppers. 'The leather is like so much brown paper, and tears off the sole after it has got one or two wettings,' wrote one soldier. Rifleman Harris, with all his cobbler's expertise, could do nothing to help them. His tools and leathers had been left on board ship.

The Walcheren fever took hold suddenly. As Harris recalled:

> At the expiration of (I think) less than a week, an awful visitation came suddenly upon us. The first I observed of it was one day as I sat in my billet, when I beheld whole parties of our Riflemen in the street shaking with a sort of ague, to such a degree that they could hardly walk; strong and fine young men who had been but a short time in the service seemed suddenly reduced in strength to infants, unable to stand upright – so great a shaking had seized upon their whole bodies from head to heel.
>
> The company I belonged to was quartered in a barn, and I quickly perceived that hardly a man there had stomach for the bread that was served out to him, or even to taste his grog, although each man had an allowance of half a pint of gin per day. In fact, I should say that, about three weeks from the day we landed, I and two others were the only individuals who could stand upon our legs.

The sickness cut through the whole army. 'You would think that the rot was amongst them were you to see how they drop off,' wrote another soldier. 'Every hour three or four are announced to have departed.' Many a soldier panicked in advance, and those who had not yet succumbed to the fever lived in hourly dread of it. Morale was not helped by the sight of wagons laden with coffins creaking through the streets, or by the news that every carpenter on the island had been set to making more of them. To reduce the panic caused by so many burials, commanders ordered that they be carried out under cover of darkness with no candles or torches, and that the fusillades that usually accompanied a soldier's farewell be dispensed with altogether. Bands playing music were also forbidden.

Some efforts were made to stop the disease from spreading, although it was far too late. Officers explained to their men about the dangers of eating unripe fruit, of sleeping out of doors, and of fishing and wading in the ditches. Observing that the local population smoked heavily – many soldiers wrote home of their surprise at observing the heavy-bottomed women sitting on their front steps enjoying a long pipe or two – officers recommended smoking at night and in the early morning. The full ration of spirits was issued, and one general counselled his battalions to drink four glasses of brandy every day – on rising, at breakfast, after dinner and in the evening. The sailors, who slept on board the ships that were anchored at a cable length or two offshore, succumbed less quickly, but in the end they too fell ill. 'Towards morning we found ourselves wrapped in that chill, blue marshy mist rising from the ground that no clothing could keep out, and that actually seems to penetrate to the inmost frame,' one soldier wrote.

Another added:

The night was clear and chill; a thin white vapour seemed to extend around as far as I could see; the only parts free from it were the sand heights. It covered the low places where we lay, and was such as you see early in the morning, before the sun is risen, but more dense. I felt very uncomfortable in it; my two hours [on watch] I thought never would expire; I could not breathe with freedom. Next morning I was in a burning fever, at times; at other times, trembling and chilled with cold: I was unfit to rise, or walk upon my feet. The surgeon told me, I had taken the country disorder. I was sent to the hospital; my disease was the same as that of which hundreds were dying.

By 27 August, barely a fortnight after the disease first broke out, 3400 men were sick. According to the words of a poem published in Edinburgh the following year:

> All fruitless victims to the cruel air,
> The climate slaughters whom the sword would spare.

Whole battalions became incapacitated, and at one stage three of them were struck off duty altogether because they had not a single man who was not in hospital. By 17 September the fever had affected 8200 soldiers, and 250 men were dying every week. Another poem, 'The Knight of Walcheren', went:

> Choice hearts! Whose welfare we should cherish,
> Not send in foreign swamps to perish.

According to Harris, the 95th Rifles had 560 men on the sick-list in early September. It became increasingly difficult for the overworked regimental surgeon to attend to all of them. The sick were so scattered that Harris and a companion named Brooks were obliged to set themselves up in a sort of 'watch-box' on the quayside, where they 'served out pills and boluses to those who came for them, which they did on the alternate days of the fever'. Whenever the bark became available they doled that out too, mixed with wine to mask the bitter taste and served in little horn cups.

To Harris, being one of the survivors started out as a matter of honour:

It was rather extraordinary that myself, and Brooks, and a man called Bowley, who had all three been at Corunna, were at this moment unattacked by the disease, and, not withstanding the awful appearance of the pest-ship we were in, I myself had little fear of it. I thought myself so hardened that it could not touch me.

It happened, however, that I stood sentinel – men being scarce – and Brooks, who was always a jolly and jeering companion, even in the very jaws of death, came past me, and offered me a lump of pudding, it being pudding-day. At that moment I felt struck by a deadly faintness, shaking all over like an aspen, and my teeth chattering in my head so that I could hardly hold my rifle.

Brooks looked at me for a moment, with the pudding in his hand, which he saw I could not take. 'Hallo,' he said, 'why, Harris, old boy, *you* are not going to begin, are you?'

I felt unable to answer him, but only muttered out as I trembled, 'For God's sake get me relieved, Brooks!'

'Hallo?' said Brooks, 'it's all up with Harris! You're catched hold of at last, old chap.'

It is difficult, nearly two centuries on, to diagnose exactly what the Walcheren fever was. 'The miserable, dirty, stinking holes some of the troops were of necessity crammed into was more shocking than it is possible to express,' wrote one brigadier after a tour of inspection of the island. Given the conditions in which the men were forced to live that summer, it seems likely that they were afflicted by several different diseases at once. A few complained of suffering from fever for three or four days at a time, with gaps of up to twenty-four hours' remission between them, and red spots appearing on their chests, stomachs and limbs, which suggests a scarlet fever; many more tell of severe stomach pains and loose bowels, which one would expect from an epidemic of dysentery or typhoid, both of which were common where water supplies were inadequate and drainage and sanitary arrangements poor. Several write about the smell of the sick, 'such as was only ever encountered in gaols'; typhus, which is spread by fleas borne by rats, was a common disease in overcrowded eighteenth-century prisons.

But the overwhelming pattern of complaint was of bouts of fever interrupted by shivering, of hard, swollen spleens – a side-effect which was famously known as 'ague-cake' – and frequent relapses. 'First a sort of listlessness pervades the mind,' one soldier wrote in his diary. 'The feet get cold, the cold then gradually ascends to the back, and in fact through the whole body; a universal icy shivering, and a chattering of the teeth, next ensue, followed by an ardent thirst.' These symptoms, together with the fact that the Peruvian bark was observed to be helpful in simple cases, points to endemic malaria, which would particularly have affected a fresh influx of people with no natural immunity to the disease. However, only *falciparum* malaria is likely to have caused death on the scale observed at Walcheren, and *falciparum*, although proven by DNA testing in Italy to have existed in Europe since Roman times, was by no means common further north.

Whatever the root cause, the abominable housing of the ill, the filth, the lack of medical attention and the overcrowding of the sick on the few ships that were available to transport them home contributed to the logistical and medical disaster that, one observer wrote, 'deluged the British empire with tears'.

The army commanders in the field were slow to react, in part because there were far too few doctors and assistants to go round, and eventually many of these also fell ill. Making an accurate assessment of what was needed took some organising, but as the number of sick men grew by the day so did the panic.

In late August, the British Surgeon-General proposed enlisting the services of London hospital surgeons and their principal pupils, called 'dressers', at suitable pay and allowances, to come to the army's aid; the authorities turned him down. By early September he was requesting additional medical assistance 'without delay'. During the following week he was demanding immediate steps to provide transports and hospital ships. Lord Castlereagh, the War Minister, declared that 'every possible exertion' had been made, but complained that it was 'difficult, if not impossible' to provide an 'adequate and immediate supply for a calamity so sudden and so extensive'.

As they waited for transport back to the 'cleaner air of England', the sick were quartered in houses, churches and warehouses, many of them burnt-out and roofless from the fighting. Bedding was in short supply: fifteen men, twelve of them sick, were discovered in a room twelve feet square, with only two blankets between them. Those being sent back to England lay on the beaches for hours on end in their filth and misery, waiting for carts to carry them down to the transports. 'The poor fellows made every effort to get on board,' Rifleman Harris recounted; 'those who were a trifle better than others crawled to the boats; many supported each other; and many were carried helpless as infants. On shipboard the aspect of affairs did not mend; the men beginning to die so fast that they committed ten or twelve to the deep in one day.'

In mid-September, as the troops at Walcheren were being laid low in their thousands by the fever, and Rifleman Harris and his friend Brooks set about dispensing medicines from their 'watch-box' on the quayside, a diligent and admirable army doctor arrived to make an inspection of the facilities.

The appointment of Sir James McGrigor was to prove a turning point in the Walcheren disaster. A Scottish regimental surgeon with a high, domed forehead and a long nose, McGrigor was a man of forthright views. He liked to make up his own mind, and was unafraid of initiating change. He, more than anyone, was aware of how crucially the army medical administration needed to be overhauled. His handling of the Walcheren disaster would catch the eye of the newly elevated Viscount Wellington, who was aware of how much his Peninsular campaign was handicapped by poor medical services. In time McGrigor became Wellington's Surgeon-General, and ultimately, as a result of his vision, energy and influence, the reformer of the Medical Department and creator of the Army Medical Corps.

When McGrigor, as Chief of Medical Staff, arrived in Zeeland to carry out his inspection in the second half of September, he was shocked at what he found. There were sick men everywhere, some of them lying on the ground without even the comfort of a blanket. 'On examining the stores, both of apothecaries and purveyors,' he wrote, 'I found them drained of many articles of the most essential description. There had been a great increase in the consumption of bark, and I found little in store.' McGrigor wrote repeatedly to the Surgeon-General in London, entreating him to hurry and send fresh supplies of the Peruvian cure. 'But before it could reach us,' he later recorded, 'we were nearly destitute of that powerful remedy.'

Knowing that stocks were within two days of being exhausted and that it would take at least a week for more to

arrive, McGrigor put out the word among the local population that he had gold, and was prepared to buy any bark that
was available. Before long he was informed that an American
ship – an 'adventurer', as he would later describe it to Parlia-
ment – had come in with supplies of champagne and claret for
the sutlers who provisioned the army. It also had on board
several chests of bark. McGrigor gave orders that the entire
shipment, 1460 pounds of it, be bought at once. It would just
tide him over until the large quantities he had ordered were
safely despatched by mail coaches to Deal and then by packet
to Walcheren.

McGrigor realised that the troops needed urgent medical
attention, and was determined to hospitalise as many of them
as possible. And as soon as was practicable, he began a whole-
sale medical evacuation. By the end of September, more than
140 ships had been commandeered to transport sick soldiers;
but arranging for the men to be moved was no easy business.
By then, nearly nine thousand of them needed help. As the
transports arrived from England, the sick were hauled up on
deck, where doses of bark were dished out in horn tumblers.
After the Channel crossing they were lifted onto the English
beaches 'like sacks of flour'. Some died as they landed; others
had to wait hours, or even days, before a hospital bed was
found for them.

In England, the hospitals were overwhelmed by the sudden
rush of new patients. For Rifleman Harris, who had found
himself among a group of the sick 'sprawling up on the fore-
castle, in a miserable state, our knapsacks and our greatcoats
over us', the conditions once he was admitted to hospital were
almost as bad as in the field. Harris was hospitalised at Hythe
on the Kent coast, and from his corner bed he saw the ward,
which held eleven men, refilled ten times, the dead being

carried out to the local graveyard. 'When I got out, and went to the churchyard to look upon their graves I saw them lying in two lines there. As they in life had been enranked, so they lay also in similar order in death.'

The British Army never forgot its terrible encounter with malaria at Walcheren, or the dangers of being caught without adequate supplies of cinchona bark. By mid-December 1809, four months after the troops first began falling ill, more than twelve thousand sick soldiers had been repatriated. Rifleman Harris was one of the thousands whose lives were saved, but he never properly recovered. The doctors attributed a large proportion of deaths to enlargement of the spleen – a classic symptom of chronic malaria – and Harris would later describe how he was 'dreadfully enlarged in the side, and for many years afterwards carried an extra paunch'.

When he was discharged from hospital Harris retired, and on a pension of sixpence a day set up as a cobbler in Richmond Street, Soho. Years later, he found himself chatting to an officer on half-pay from the Oxfordshire Light Infantry, who wrote down Harris's army recollections.

Meanwhile, many of the soldiers who followed Wellington's call to Spain, where McGrigor served as Surgeon-General, could be recognised by their Walcheren 'waistcoats', a flannel vest with long sleeves made 'large enough to wrap around the breast', and the little bottle of Peruvian bark they carried in their holsters instead of a pistol.

The *Statistical Reports on Sickness, Mortality and Invaliding Among the Troops in Western Africa* that is housed today in the rare books section of the British Library is bound in navy-coloured leather, and is so little consulted that its pages bristle when opened, giving off a faint smell like that of an old powdered wig. Yet when it first appeared in 1840, the book was revolutionary. Here, compiled for the first time, was a detailed and accurate record of the health of every British soldier serving abroad. The move was one of the many innovations introduced by Sir James McGrigor when he began his reform of the Army Medical Department after the catastrophic Walcheren expedition. It would have a profound influence, not just on the way the army dealt with the prevention and treatment of disease among soldiers, but on the expansion of the British Empire.

Nowhere was this more apparent than in West Africa. Soldiers preparing for a posting to Bight of Benin knew that at virtually any spot on the coast they could be certain that within a year or two they would have buried at least half the regiment.

In 1832, the first expedition to attempt a complete exploration of the River Niger departed from Britain carrying a pitiable medicine chest that contained little more than calomel, Epsom salts and citric acid. The expedition leaders quickly fell ill with malaria, and many of the members died. Sulphate of quinine, a soluble salt isolated from the natural quinine found in the bark, was included 'as a strengthener after fever', but the niggardly supply of just four ounces did nothing for them.

Nor were other parts of the coast any better. Sierra Leone, which had been established as a home for liberated slaves, was known as 'a pestiferous charnel house'. In the capital Freetown, the African remittent fever, as malaria was called there, was so serious that two years after the first Niger expedition, the city fathers seriously considered building a wall up to twenty miles long and thirty feet high to stem the invasive miasma that they believed was the cause of the fever.

The Gold Coast, next door, was said to be 'the most unfriendly to men of any country on the face of the globe'. At Fernando Po, an island off the west coast of Africa, the standing orders for labour forces were 'Gang No 1 to be employed digging graves as usual. Gang No 2 making coffins until further notice.' As for the Bight of Benin, it inspired this sinister sea shanty:

> Beware and take care of the Bight of Benin,
> There's one comes out for forty goes in.

In 1840, the first statistics on the health of the troops posted to West Africa revealed that for every thousand soldiers there were 2978 admissions to hospital. That meant that every man could expect to be hospitalised three times a year. Worse still, 483 of those thousand soldiers would die before they could be discharged from hospital.

Many succumbed in the Gambia, on the most western reaches of the coastal bulge, 'which proved,' the report said, 'the grave of almost every European sent there'. The first detachment of 108 British troops arrived at the Jamestown garrison, at the mouth of the Gambia river, at the end of May 1825, just as the rains started. By 21 September, only twenty-one were still alive, seventy-four having died of malaria. Another ninety-one were kept at sea, aboard a transport named the *Surrey*. They did not lose a man. But the deaths of their comrades made room for them onshore. Between the end of September and 21 December, sixty-one more had died. The force having now been reduced by deaths to thirty-nine, most of whom were in the final stages of disease, another body of two hundred men was sent in their place. After less than three months, half that number were in their graves, eighty-five of them as a result of malaria. 'The disease,' the report concludes, was 'of the most malignant form, and even in the event of recovery, [it left] the patient with a shattered constitution for life.'

One morning in early May 1854, William Balfour Baikie turned into the John Laird yard at Birkenhead and caught sight of the ship that would carry him up the Niger, the *Pleiad*. As a graduate of Edinburgh's prestigious school of medicine and a surgeon in the Royal Navy, Baikie had read the statistical reports on the health of soldiers, and knew all about the risks involved in an expedition to West Africa.

Just 105 feet in length, the *Pleiad* was the first exploring ship to be designed specifically so that she would not have to take on wood from the African coast, a known source of fever, as fuel. She had five staterooms for officers, all elegantly fitted out with mahogany tables, green morocco leather sofas and a bronzed chandelier, and their own bathrooms. An ample library completed the suite of furniture. The dozen sabres that hung on either side of the main door of the *Pleiad*'s library added to her air of buccaneering confidence.

The shipwright MacGregor Laird had agreed to the Admiralty's conditions, that he should build the steamer and pay all the expenses of her maiden voyage to West Africa for a fee totalling £5000, because he had a hunch that despite its reputation as a hell hole, the region had great commercial potential; and Laird was a man who liked to be in at the start.

Looking about him, Dr Baikie knew he was right to have taken his former supervisor at the Haslar hospital Sir Roderick Murchison's advice to volunteer for the next Niger expedition. As a poor Orkney boy, the eighth of thirteen children, Baikie had to take his opportunities when they came. He had struck Murchison as a potential explorer because of his medical training and his enthusiasm for field botany. The *Pleiad*, which would be the expedition's home, left Dublin after her sea trials on 20 May. Baikie and most of the other expedition members followed four days later in another packet, the *Forerunner*.

The Niger did not hold the same romance as the Nile in the eyes of the Victorians. The river would never have a Livingstone to capture the imagination of the British public, although the romantic Mungo Park, with his Byronic locks and mysterious death in 1806, came close. Nor were its explorers lauded, as the later Nile generation was to be, in the *Illustrated London News* and asked to speak at the Royal Geographical Society. But in the mid-nineteenth century, the Niger was about to overtake the Nile in commercial importance. After centuries of disagreement over whether the river was a minor tributary of the Nile or a part of the Gambia, the Niger had recently been proved to be a giant waterway in its own right, earning its local name as the 'River of Rivers'.

Rising in the hot flatlands of northern Guinea, it flows in a gigantic semi-circle, 2600 miles long, before shattering into innumerable rivulets in the delta that spills into the Bight of Benin. On the way, the muddy grey whale of the Niger passes through regions as disparate as the white desert sands of central Mali and the verdant borderlands of Nigeria's frontier with Niger. Its massive eastern tributary flows through the thick green forests of central Cameroon which would provide Edgar Rice Burroughs with the setting for his tales of the ape man, Tarzan.

The fact that the Niger was one of the unhealthiest regions on earth did not deter explorers, missionaries and traders from trying to penetrate it as far as they could. Spurred in part by the Lairds and other European traders, the Niger was about to become the backbone of a huge trade in palm oil that would reach as far as the central Sahara, and lead to the division of West Africa between Britain, France and Germany. The naval surgeon Dr Baikie, with his near-obsessive prescriptions of quinine as a prophylactic against malaria, would ease the way.

The *Forerunner* passed through the Bay of Biscay and across to Madeira, where the passengers went briefly ashore to buy fresh bunches of cherries. From there they headed south again, around the West African bulge. Shoals of flying fish leapt out of the water, some landing on deck, where they were scooped up by the ship's cook. Baikie paid them little attention, for he was too busy reading. Sitting in his stateroom, or, when the weather was not too rough, out on deck, he devoured the accounts of earlier travellers to the region, including Mungo Park, like Baikie a Scot and a naval surgeon, who had fallen sick with 'the fever' within less than a month of landing in West Africa after 'imprudently exposing myself to the night dew, in observing an eclipse of the moon'. Park reported that the next day he found himself 'attacked with a smart fever and delirium'; his recovery was very slow, and he was prone to relapses. In November 1805, Park's expedition surgeon, John Martyn, wrote in his journal: 'Thunder, death and lightning: – the devil to pay: lost by disease Mr Scott, two sailors, four carpenters, and thirty-one members of the Royal African Corps, which reduces our numbers to seven.'

The mortality rate of the earliest expeditions to the Niger was frightening. In the course of Park's second journey in 1805, of forty-four Europeans who accompanied him from Gambia, thirty-nine died. In 1833, MacGregor Laird, who commissioned the *Pleiad* and was the son of the principal Liverpool shipowner trading with Africa, had returned with only eight of the forty men with whom he had set out.

Such hardship would have been familiar too, Baikie read, to the cheerful Lander brothers, two young indentured servants from Truro who found themselves leaderless when their master, Hugh Clapperton, died in Sokoto in April 1827, more than three months' journey from the sea, before having achieved his goal, the discovery of the source of the Niger.

Like Mungo Park before them, and indeed Clapperton too, the Landers, especially John, the pensive and naturally studious younger of the two brothers, regularly succumbed to the African remittent fever. Dosing themselves with little more than soda powder and calomel, it is not surprising that they so often began the journal they wrote together with the words: 'I found my brother in a high fever this morning.'

Sitting one lonely evening outside a grass hut assigned to him by a friendly local chief who also sent along a plate of little cakes made of pounded corn, the normally cheery Richard wrote of his younger brother: 'Towards evening he became worse, and I expected every moment to be his last. The unhappy fate of my late master, Captain Clapperton, came forcibly to my mind. I had followed him into this country where he perished; I had attended him in his parting moments; I had performed the last mournful office for him which our nature requires, and the thought that I should have to go through the same sad ceremonies for my brother overwhelmed me with grief.'

Despite his brother's fears, John did not die that night. But the Landers' experience, and that of the travellers who had preceded them up the Niger, made Baikie all the more resolute about guarding the health of the passengers and crew of the *Pleiad*. As the only medical officer on board, the responsibility would fall to him. He was determined that the expedition would adhere to a strict medical and dietary regime. All its members would have to take daily doses of quinine, which Baikie had seen used in the West Indies and in the eastern Mediterranean, and which had had such success in treating tertian ague and intermittent fevers in the marshy lands of England and continental Europe.

By the time the *Forerunner* caught up with the *Pleiad* near Fernando Po after a month at sea, though, Baikie had more immediate problems on his hands. The leader of the expedition, John Beecroft, the British Consul of the Bight of Benin and Biafra, had fallen ill and soon died. Baikie, as the highest-ranking government officer left on board, had no alternative but to take charge himself. He was not yet thirty, and had never set foot in Africa.

Among the many books and journals he had read on the month-long journey out to West Africa was the medical journal of an expedition nearly twenty years before which had been led by another Scottish naval surgeon, James Ormiston MacWilliam. MacWilliam's record of the endless recurrent fevers with which the passengers and crew fell ill, and the harrowing accounts of their treatment, cannot fail to have impressed Baikie. Like so many doctors before him, MacWilliam favoured castor oil, enemas, mercury, blood-letting and blistering, especially on the back of the neck. Such treatments were difficult enough to administer even in a European sickroom, and one wonders how many patients succumbed to infection after a bloodletting, however well intentioned, in the thick humidity of the West African coast.

What is most interesting in MacWilliam's account is the importance that he places on prescribing quinine. Of his sixty-two crewmen, at least fifty-five fell ill with what he calls African remittent fever. He dosed them all with quinine. Although by the end of the expedition sixteen had died, that was half the ratio that the British Army had come to expect among soldiers posted to the West African coast. 'No medicine was found so efficacious,' MacWilliam wrote in his medical memoir of the Niger expedition, 'as quinine.'

As the *Pleiad* began its journey up the Niger delta, Baikie was so taken up with the novelty of Africa – the lush vegetation that tumbled into the grey river, the roars of animals he could not see on the forested riverbank, the piles of ivory and hippo teeth to be bought by day, and the sight of the Southern Cross in the sky by night – that he paid little attention to his growing dislike of the ship's sailing-master. By all accounts, Thomas Taylor could not have been more different from the serious-minded Baikie. He drank on watch, bullied the *Pleiad*'s African crew and forced them to work on the Sabbath, which caused the God-fearing Baikie to purse his lips in disapproval.

For as long as they needed to, the two men tolerated each other. But in early October, when the *Pleiad* had been on the river for nearly three months and had penetrated as far as Dulti, further than any European transport had been before, Taylor's inattention caused the ship to run aground. 'The real cause was too apparent,' Baikie wrote. Two weeks later, despite having been severely reprimanded, Taylor ran the *Pleiad* aground a second time. But it was the discovery that the crew were beginning to show symptoms of scurvy that led Baikie to relieve Taylor of his command. That the *Pleiad*'s crew was suffering from a deficiency of vitamin C, which had been almost unheard of in the Royal Navy in the eighty years since Captain Cook proved that the simple introduction of lemon juice into the diet was enough to prevent it, was the direct result of Taylor's tight-fistedness in victualling the ship with rice and little else. With the help of Daniel May, the second master, Baikie took over the ship himself, and immediately put ashore to buy supplies of meat and fresh fruit.

Baikie's assumption of responsibility for the *Pleiad* and her passengers and crew did much to foster the young doctor's capacity for decisive leadership. It no doubt also allowed him to be extra vigilant in enforcing the medical regime he had planned. No one on the *Pleiad* was going to fail to guard against the fever by taking quinine.

At the end of the expedition Baikie returned home and wrote a book about his travels, which went by the cumbersome title of *A Narrative of An Exploring Voyage up the Rivers Kwo'ra and Bi'nue (commonly known as the Niger and the Tsadda)*, published in 1856. Although it lacks the verve of the later accounts of Henry Morton Stanley, who was as much a show-off as Baikie was a puritan, or the fervent zeal of David Livingstone, the energy and enthusiasm of Baikie's book carries the reader with him.

Even more fascinating is the forty-page treatise that he wrote about his success with quinine, which is lodged in the British Library but has never been published. Baikie had reason to be pleased with the care he bestowed on his fellow travellers, and they in turn were more than grateful to him. Of the twelve Europeans and fifty-four Africans aboard the *Pleiad*, one was murdered and another drowned when he fell into the Niger. Not one died of the fever. 'While up the Niger in 1854 I had ample opportunity for testing this virtue [of regular doses of quinine], and I most unhesitatingly record my belief in its existence. While in swampy districts it was regularly administered, and its use continued for about a fortnight afterwards, and among twelve Europeans hardly any sickness occurred during our stay of four months in what has hitherto been considered a most unhealthy river.'

'One of the greatest improvements in tropical medical practice of late years,' wrote Baikie in his neat copperplate after a long preamble on the complicated nature of the various remittent fevers, 'has been the employment of quinine not as a curative agent but as a prophylactic or preventive.' Baikie had insisted that the passengers and crew of the *Pleiad* take a daily dose of six or eight grains of quinine, half in the morning and half in the evening, 'dissolved in sherry,' he insisted, 'as in the Naval service'.

Baikie had seen quinine used to great effect in the West Indies and in Smyrna in Asia Minor, where malaria was particularly rife. His faith in it was based on observation and experience, for he knew nothing about the root causes of malaria, or how quinine cured it. In his attitude towards medicine he was far less like the Europeans and Americans who, over the centuries, have arrived in Africa with fixed ideas and an arrogant insistence that things were not done well unless they were done *their* way, and more like the independently-minded settlers who used their inventiveness to devise ways of working – planting, healing, repairing – that best suited the environment in which they found themselves.

That attitude did not stop Baikie, however, from devising elaborate theories about the causes of malaria and how to be rid of it. Although he wrote often of being attacked by insects, and even, when he was particularly vexed, by 'legions of bloodthirsty mosquitoes', he never suspected them to be carriers of the disease. Instead, he subscribed to the common notion that malaria was something that grew naturally out of the vegetation in marshy areas. 'Walking along pathways through the brushwood,' wrote one of his fellow travellers on the *Pleiad*, 'and far away from what we are apt to consider as the indispensibles of decomposition, we occasionally become conscious of a "steamy vapour", which flies up with intensity sufficient to bring on an attack of ague (with persons predisposed to it) in the course of a few hours.' The free circulation of oxygen, the German chemist Justus von Liebig wrote not long before Baikie left England, was 'the enemy and opponent of all contagion and miasma'. While Liebig's pronouncement had made sense in Europe, in the midst of the Niger delta, with as much oxygen as was imaginable, the fever and 'miasma' or 'steamy vapours' that caused fever were still ever-present.

Baikie took up another theory set out in MacWilliam's memoirs, namely that malaria was caused either by 'submarine volcanic action', or the 'reaction of vegetable matter upon the saline contents of the water'. Decaying vegetable matter, Baikie had been told by a distinguished chemist named Professor Daniel, 'abstracts the oxygen from the sulphate of soda contained in sea water, and a sulphured of sodium is formed. This again, acting upon the water, decomposes it, and sulphuretted hydrogen gas is one of the products of the decomposition.' Without explaining quite what this gas might be, the inference was nonetheless clear: that sulphuretted hydrogen gas was the basis of malaria.

Baikie set one of the expedition members to collecting water samples. Eight bottles of water from different parts of the coast were sent back to Professor Daniel for analysis; none of them proved to contain any more sulphuretted hydrogen gas than was present in a comparative sample from Harrogate.

However quaint Baikie's theories seem to us now, his faith in quinine proved enduring. That none of the travellers aboard the *Pleiad* died of fever in the course of the four-month expedition is remarkable. In the context of what came later, this expedition marked a turning point in the exploration and colonisation of Africa.

Baikie's success was not limited to his prevention of malaria. He returned to West Africa at the head of a second expedition, and was soon running one of the most successful trading markets in the region. By this time he had taken to wearing a turban, a long cotton shirt called a *tobe*, and baggy trousers like a *zouave*. He had an African mistress and had fathered several children, for whom he translated the Book of Genesis into Hausa. But Baikie would grow bitter and disillusioned in Africa. In 1859 he wrote to his patron Sir Roderick Murchison, by then the President of the Royal Geographical Society: 'I am entirely exhausted. I have not been in anything like a bed for more than two years. I have had fever on more than 120 separate occasions, not reckoning mere recurrences.' He died of dysentery on his way home to England in 1864, having got no further than Sierra Leone.

Seven years after Baikie's first expedition, and as a direct result of its success in treating the fever, the British government annexed the trading centre of Lagos and the lagoon behind it, thus establishing the beginnings of the first British colony on the West African coast. There was such a rapid expansion in the palm oil trade that MacGregor Laird was encouraged to form the African Steamship Company to cope with it. His efforts led the British government to become more interested in the interior of West Africa, and eventually to support and protect a steamship route up the lower part of the Niger.

Within twenty years, France and Germany too became directly involved in Africa. It is not hard, looking back, to see the straight line that leads from Baikie's breakthrough with quinine to the Berlin conference of 1884 that led to the partition of the continent and the takeover of so much of it by the major imperial powers. While it was still not known what caused malaria, its cure, quinine, which had been such a totem to religious power in the seventeenth century, was beginning to turn, as the world changed, into a symbol of the growing power of commerce and exploration.

To Explore and to War –
From America to Panama

Ma called Pa, and he came in.
  'Charles, do look at the girls,' she said. 'I do believe
they are sick.'
  'Well, I don't feel any too well myself,' said Pa. 'First
I'm hot and then I'm cold, and I ache all over. Is that the
way you feel, girls? Do your very bones ache?'

LAURA INGALLS WILDER, *The Little House*
on the Prairie (1935)

'Death was constantly gathering its harvest all about me.'

PHILIPPE BUNAU-VARILLA, during the malaria and
yellow fever epidemics in Panama City, 1885

When Laura Ingalls Wilder set out to write about her family,
how her Pa and Ma, Mary and Laura and Baby Carrie left
their little house in the Big Woods of Wisconsin and headed
out west, she knew she'd be writing about more than just ox
wagons and the wide open plains. By the time they'd got to
Kansas, the family had been through rising creeks and barren
valleys, met a wolf pack, a pair of Indians wearing skunk
skins, a Tennessee wildcat and a herd of Texas longhorns. But
it wasn't until the blackberries were ripe, at the end of the
summer, that they encountered the fever.

Laura did not feel very well. One day she felt cold even in the hot sunshine, and she could not get warm by the fire . . . Ma stopped her work and asked, 'Where do you ache?'

Laura didn't exactly know. She said: 'I just ache. My legs ache.' . . .

Ma and Pa looked a long time at each other and Ma said, 'The place for you girls is in bed.'

It was so queer to be put to bed in the daytime, and Laura was so hot that everything seemed wavering . . .

Something dwindled slowly, smaller and smaller, till it was tinier than the tiniest thing. Then slowly it swelled till it was larger than anything could be. Two voices jabbered faster and faster, then a slow voice drawled more slowly than Laura could bear. There were no words only voices.

Mary was hot in the bed beside her. Mary threw off the covers, and Laura cried because she was so cold. Then she was burning up, and Pa's hand shook the cup of water. Water spilled down her neck. The tin cup rattled against her teeth till she could hardly drink. Then Ma tucked in the covers and Ma's hand burned against Laura's cheek . . .

She lay and watched Mrs Scott tidy the house and give medicine to Pa and Ma and Mary. Then it was Laura's turn. She opened her mouth, and Mrs Scott poured a dreadful bitterness out of a small folded paper on to Laura's tongue. Laura drank water and swallowed and swallowed and drank again. She could swallow the powder but she couldn't swallow the bitterness . . .

Mrs Scott said that all the settlers, up and down the creek, had fever 'n' ague. There were not enough well people to take care of the sick, and she had been going from house to house, working night and day.

'It's a wonder you ever lived through,' she said. 'All of you down at once.'. . .

No one knew, in those days, that fever 'n' ague was malaria, and that some mosquitoes give it to people when they bite them.

Towards the end of summer, when the earth was baked hot and the rivers and creeks were drying up, and they'd been so long without rain to sluice them clean, was the time when malaria used to take hold in the United States. Early travellers complained frequently of suffering from intermittent fevers, agues and fluxes, though only occasionally did they report in detail the presence of malaria's more particular symptoms – the fever interspersed with shivering and the swollen, painful spleen.

Like the southern coast of England, it seemed that the eastern coastal areas of North America were prone only to occasional bouts of malaria. But by the end of the seventeenth century the seasonal outbreaks of 'fever 'n' ague' had become full-blown epidemics, spread in part by a growing population that had no natural immunity to the disease.

European ships, especially those that sailed in the summer months when malaria was at its most prevalent along the Kent coast, in the Fenlands and throughout East Anglia, were probably the earliest carriers of mosquitoes harbouring the *Plasmodium vivax* and *Plasmodium malariae* parasites to the New World. Later, slave ships from Guinea, which first reached the rice-growing areas of the South Carolina coast in the 1680s, quite probably brought the more deadly *falciparum* strain with them.

As in Peru, it seems unlikely that malaria existed in North America before the arrival of the white man. By 1610, it may have been present to greet the Governor of Virginia Baron de la Warr, who 'arriv[ed] in Jamestowne [and] . . . was welcomed by a hot and violent ague'. The colonists' letters, journals and diaries were soon speaking constantly of enduring the 'seasonings' and 'fevers'. Laura's 'fever 'n' ague' followed frontiersmen who landed in New England and made their way to Georgia and all the way to California. The best and most fertile acres, those in the flatlands of rivers, creeks and streams, held the greatest appeal to settlers, and it was natural that those were the areas that the early travellers chose to take over first; but they were also the places where the *Anopheles* mosquito flourished. From Virginia, where the Jamestown settlement nearly failed completely because of disease, the Reverend John Clayton wrote in 1687 that intermittent fever was first among the illnesses that attacked the English settlers. Another colonist reported that his sister, having suffered from two or three fits of fever and ague, was now well over the 'seasoning'. A Scottish settler in Tidewater, Virginia, wrote in 1723 that he suffered greatly from 'fever and Ague wch is a very violent distemper here', adding rather bitterly that the place was 'only good for doctors and ministers'.

In 1735, the wealthy Virginia landowner William Byrd wrote to the Governor of North Carolina, who was suffering from malaria, warning him to take repeated dosages of the 'Bark', or else his 'feaver' would 'as surely return as speech to a silent woman'. Thomas Salmon, describing Virginia in 1738, wrung his hands in anxiety that 'Fevers and Agues, the Gripes and Fluxes are the most common Distempers.' Similarly, Philip V. Fithian recorded in his journal on 12 August 1774: 'The conversation at Table was on the Disorders which seem to be growing epidemical, Fevers,

Agues, Fluxes – A gloomy train.' And a new settler was even more anxious when he arrived in 1785, to find that the colony was 'Esteem'd the most sickly Province this Except Georgia & South Carolina. Fevers and Agues, Plurises, Bilious Fevers rage Terribly . . . it appears to me like a general plague.' Seven of his roommates, he continued, were 'raging out of the senses' with 'Agues and Fevers'.

New England troops fighting for the first time in the South in the Revolutionary War were particularly susceptible. As early as 1776, the Continental Congress ordered that three hundred pounds of Peruvian bark be sent to the Southern Department to treat them. By the time the war was well under way, and Dr Craik was treating George Washington's tertian fever in August and September 1786 with eight doses of red bark at a time, malaria was about to extend its hold even further westwards.

The first settlers to cross the Appalachian Mountains expressed their amazement at the beauty and apparent healthiness of the Ohio Valley. Within a short time, though, they were commenting upon the appearance of 'fevers and agues', and soon travellers were noting the sallow skin and anaemic appearance of the residents. With the restrictions barring the westward expansion of settlement removed at the close of the Revolution, an unparalleled movement of indigenous and immigrant populations began which would last for well over a century. Everywhere the new emigrants went – over the Allegheny Mountains, across the Great Lakes, along the Grand River, in the vicinity of the Rideau canal and into upper Canada – malaria would follow. In 1793 a Mrs Jarvis, who lived in Niagara, wrote to her father in England imploring him to bring 'plenty of bark, for the children are all ill with fever'.

Since the rivers were still the principal routes of travel across the United States, a huge proportion of this migration was obliged to pass through intensely malarious regions, from which it would spread the disease further afield. In this way malaria was carried up the river valleys of the Atlantic coast, across the Appalachian divide into western New York State and Pennsylvania, and into the Ohio Valley. In 1810, one writer warned all travellers who proposed to travel down the Ohio and Mississippi rivers to provide themselves with mosquito curtains, otherwise they could not reckon on a single night's undisturbed repose during the spring, summer and autumn. Later, writing of the river settlements in the Mississippi territory, he says: 'on the subsistence of the waters the sickly season commences and lasts with little variation from July to October. The driest seasons are the most unhealthy. The prevailing malady is a fever of the intermittent species, sometimes accompanied by ague, and sometimes not. It is rarely fatal in itself, but its consequences are dreadful, as it frequently lasts for five or six months in defiance of medicine, and leaving the patient in so relaxed and debilitated a state that he often never regains the strength he had lost.'

From the southern Atlantic states, the fever was soon to be found in Kentucky, Tennessee, Alabama, Mississippi, Arkansas and northern Louisiana. The early overland immigrants to the west coast also carried the disease with them, and it took hold on the lower Columbia River and in the Sacramento–San Joaquin river valleys. By 1850, practically the entire United States constituted one vast expanse of malarious country, except for Maine, the northern portions of Wisconsin and Minnesota, the Appalachian highlands, the cool, arid north-west plains, the Rocky Mountain area, the western desert and the Sierra ranges.

In many regions the disease was mild, and occurred only in the warm summer months. The *Plasmodium vivax* and *malariae* that came, probably, from Europe were the most common and, fortunately, the least serious of the malaria parasites, and usually died out in the winter months unless the weather was unusually mild. But in an extensive southeastern triangle from Baltimore to central Florida and westward to the state of Mississippi, as well as in the areas drained by the Ohio and Mississippi rivers and in eastern Texas, where malaria persisted all year round, the *falciparum* strain was more common. An English proverb of the time said, 'They who want to die quickly go to Carolina,' while in 1737 a German observer told his readers: 'Carolina is in the spring a paradise, in the summer a hell, and in the autumn a hospital.'

Many of the Africans who had been shipped to the South as slaves suffered from sickle-cell anaemia, an inherited blood condition common among Africans and some peoples living around the Mediterranean that has the unusual benefit of offering some natural resistance to *falciparum* malaria, as it stops the malaria parasite from reproducing in them. The result was that slaves seemed less likely to catch malaria; those most prone to it were new immigrants and white northerners visiting the South, a fact that became obvious when the Union forces were on the offensive during the Civil War.

The conflict between the North and the South was a watershed, not just politically and militarily – the triumph of the Union represented that of the industrial over the agrarian – but medically as well. The American Civil War was the last great conflict to be fought without any knowledge of germs and how they spread disease. Within a few years perspicacious men such as Louis Pasteur, Joseph Lister and Robert

Koch would shed fresh light on the nature of microbial infection. But for the hapless participants of that early war, sickness and death from the commonest causes still remained a mystery, striking savagely at both sides. As far as medicine was concerned, the American Civil War was like the Middle Ages before the Renaissance – as one observer described it, 'the sepulchre of unscientific medical thinking and a monument to the awesome power of disease'.

In the 1860s epidemic infection still ran riot, unchecked by antiseptics, quarantine or even simple decent cleanliness. Measles, smallpox and typhoid were common. Of all the infectious diseases, only smallpox had been proved to be preventable by vaccination, and even that was still the subject of debate. In some circles, especially in Philadelphia and Missouri, malaria was known to be alleviated by quinine, but no two doctors agreed either on dosage or on timing.

Indeed, at the time of the Civil War so little was known about preventive medicine of even the most basic kind that little effort was made to keep military camps hygienic. Training camps brought together huge numbers of susceptible men, many of them young and of rural origins, who may not previously have been exposed to childhood illnesses. Measles and mumps took their toll. Soldiers in the field preferred to relieve themselves close by their tents rather than at any distance for fear of being caught 'with their pants down'. Water was thus almost always contaminated: 'I could hardly get my horse to drink it,' complained one Texas surgeon at the start of the war.

As time wore on, men fed on poorly preserved salt pork, hardtack (stale biscuits) and heavily boiled vegetables – what the troops called 'desecrated greens' – suffered badly from malnutrition, which did little to aid their recovery from disease. And the wounds received in battle were virtually welcoming posts for gangrene and fatal bacteria. The .57-calibre minié balls fired by most small arms were of such low velocity that they usually introduced filthy bits of cloth from the soldiers' tattered uniforms into wounds, but offered no sterilisation. Any bone wound was invariably dealt with by amputation, while nearly two-thirds of those suffering intestinal injuries died. One Civil War general, describing a typical surgical routine, wrote: 'There stood the surgeons, their sleeves rolled up to their elbows, their bare arms as well as their linen aprons smeared with blood, their knives not seldom held between their teeth, while they were helping a patient on or off the table . . . The surgeon snatched his knife from between his teeth . . . wiped it rapidly once or twice across his bloodstained apron, and the cutting began.' Mortality from battlefield infection was horrendous, with almost 275,000 Union soldiers dying out of a total of 2.2 million, according to figures collated by the Surgeon-General shortly after the war, and 164,000 out of 750,000 soldiers in the Confederate army.

For every death from battle wounds, there were two from disease. Although this ratio represented some improvement over earlier conflicts – in the Crimean War it had been three to one; in the Mexican–American War seven to one; and for the British forces in the Napoleonic Wars as high as eight to one – death by disease during the American Civil War was still so common that the dead were known, after the blue and grey forces, as the 'Third Army'. The most common causes were pneumonia, influenza and bronchitis. But epidemics of malaria also took their toll, especially during the early years.

From 1861 on, Union attempts to establish a bridgehead along the Atlantic coast were stopped largely because of outbreaks of malaria, typhoid and yellow fever. In 1862 General Henry Halleck's forces in the west, which totalled more than 100,000 men, were unable to overcome the threat of malaria to take control of the Mississippi at Corinth. Soon after, another attempt to take the Mississippi at Vicksburg was also thwarted by the disease. As a result, it would take more than a year before the Union forces succeeded in taking Vicksburg.

The town lay about four hundred miles above New Orleans and the same distance below Memphis, on a high bluff on the east bank of the river. The surrounding low countryside was cut by bayous that were prone to flooding during high water. Much of the land around was known then as the DeSota Swamp. During most of the campaign, which began in May 1862, the weather was hot and dry, but there was much green-scummed standing water left behind when a recent bout of flooding had subsided, the perfect conditions for mosquitoes to breed. Plantation owners in the area reported that their families and slaves suffered unusually severely from malaria that summer, and it is likely that the local population was responsible for the rapid spread of the disease. Vicksburg then had the second highest mortality rate from malaria among twenty-three Southern cities, worse even than New Orleans. 'This is certainly the most unhealthy place I have seen,' wrote Assistant Surgeon Junius N. Bragg of the 11th Arkansas Regiment, who suffered for three years from recurrent malaria, while General Halleck wrote in June 1862, less than a month before he was relieved of his command: 'If we follow the enemy into the swamps of the Mississippi there can be no doubt that our army will be disabled by disease.'

How right he was. The eventual seizure of Vicksburg, and with it control of the Mississippi River, marked the end of the possibility of a Confederate victory, but it came at enormous cost. The further south the Union forces advanced, the more common it was for their soldiers to die of malaria. Many of them, especially those from Kentucky and north Connecticut, had never been so exposed to mosquitoes. Within days they began falling sick. The 1st Kentucky Brigade lost a third of its strength within a month; another regiment was reduced from nine hundred men to 197 in an even shorter time. The company that replaced it started with a hundred men in mid-June; by the end of July there were only three left. Virtually every regiment at Vicksburg that summer reported that it was short of rations, and crucially, also of quinine. When supplies were exhausted, requisitions sent to New Orleans were not honoured by the medical director because of 'irregularity'.

In the Union army alone, 595,544 ounces of quinine sulphate and 518,957 ounces of fluid cinchona extract were issued by the authorities. The exactness of the figures, down to the last ounce, says much for the efficiency of the Union quartermasters. But down the line, supplies were nowhere nearly so well organised. Regiment after regiment in the early years of the war began campaigns in the summer months, rather than during the frost when the danger of catching malaria would have been far lower. From the Carolina coast, from Corinth, Richmond, Vicksburg, Chickahominy and Baton Rouge, came reports of camp doctors trying to treat malaria with capsicum, nitric acid or boiled bark from willow or dogwood trees; or, failing that, with just straight whiskey, because the quinine had run out and much-need replenishments had failed to arrive.

By the end of the Civil War, the three-volume *Medical and Surgical History of the War of the Rebellion* (1861–65) reported, the Union army had diagnosed more than 1.3 million cases of malaria. The disease was largely untreated, and by the end of the war had killed more than ten thousand men.

The higgledy-piggledy manner in which soldiers on both sides of the war were supplied with medicine can be blamed on many things: medical ignorance, diagnostic disagreements, bad communications and long distances to name but a few. That so many doctors still disagreed about prescription and treatment, especially of remittent fever, is perhaps the most surprising factor, for the healing properties of quinine had been known for more than two centuries. Among American doctors, no one believed in its qualities quite so much as the great-great-great grandfather of Ginger Rogers, the pioneering Dr John Sappington, whose Sappington's Fever Pills became one of the most successful patent medicines in America in the early nineteenth century.

Dr Sappington, who was just six weeks old when America gained its independence in 1776, came from a pioneering medical family. He was born in Maryland, on Chesapeake Bay, but by 1785 his father had moved the family from that genteel environment westward to Nashville, Tennessee. Young John began his medical studies as an apprentice to his father at an early age. When the time came for more formal learning, he attended a term's lectures at the Philadelphia Medical College, but left before he had graduated, disheartened perhaps by the harsh winters and too much theory. In 1817, he followed his father's example and moved his seven children from Nashville westward to the frontier of mid-Missouri, about two hundred miles up the Missouri River from St Louis.

Although the region was sparsely populated, within five years he had built up a sizeable practice. As a frontier physician, John Sappington was a well-known pioneering type. With his medicine case in his saddlebags, he rode the narrow trails to any household where someone ill awaited him. It might be in the heat of summer, or in the dead of winter when his horse was forced to kick aside the snow; it might be in the night, when his body called for sleep, or on the day after such a night; it might be to attend the well-to-do, or some poor person who could not be expected to pay. Sappington's charges were typical of the times. His ledger books show numerous entries such as 'to visit and riding 10 miles, $3.50', and occasionally 'to riding 65 miles and three days and nights attendance, $50'. Even more frequent are the notations of payment in salt, dry beans, work, 'one hind of beef', bacon, corn, wheat or ferriage, and on one occasion, two barrels of whiskey.

Dr Sappington became renowned for treating malaria with powdered cinchona bark, avoiding the dreadful practices of bloodletting, purging and the huge doses of harsh drugs that were more typically employed. Indeed, his reputation was such that families with children sought to settle close by his farm at Arrow Rock so that he would be close at hand for the inevitable malaria attacks in the summer and autumn. In later years one visitor, Captain J.A. Pritchard, described Sappington as 'a very adroit and singularly eccentric character – jocular and lively and rather quisical, possessing a high degree of hospitality and gentlemanly demeanour – He is a large fine looking man about 6 feet high and looks to be about 70 years of age with heavy suit of hair and it white as snow – his beard as white as his head and hung to his breast.'

When Sappington was learning medicine, it was still commonly believed that fever was the result of an irritation or excitement. Thus the first step in its treatment, many physicians believed, was to 'calm' the patient. This would generally be done by bleeding or by repeated uses of purgatives and emetics such as calomel, or even pepper, tobacco, rhubarb and lobelia powder, or a combination of these methods. Quinine, which had been introduced to America in about 1745, had never been very popular in any form – due, one authority says, to its early abuse by quacks. The prevailing view was that it was a stimulant which should properly be administered in the absence of fever, and then only after a period of preparation by all the bloodletting and purging that the patient could stand.

What Captain Pritchard took to be eccentricity in Dr Sappington might rather have been evidence of an independent spirit. Sappington differed fundamentally from other doctors. He might bleed if a patient seemed 'overly full-blooded', but only moderately and very rarely. He held to the then unusual view that the blood, with its load of oxygen and 'food substances', was needed for vitality. He was unusual for his time too in advocating cool water for those suffering from fever, reasoning that it replaced liquids lost through perspiration, as well as being very pleasant to patients. He was also among the first doctors to use quinine not just as a cure, but also as a preventive.

In later life, Sappington recalled that he first learned about cinchona bark a short time after finishing his medical apprenticeship when he came across a pamphlet recounting the history of its discovery in Peru, or at least the fictitious version involving the Countess of Chinchón, and of its use in fevers. He began by experimenting on himself with one-ounce doses (equivalent to 10–12 grains) of sulphate of quinine while he was well. Emboldened by the fact that none of the side effects he'd been warned of seemed to appear and that his pulse and temperature remained normal, he began trying the drug out on his feverish patients, though not without some anxiety.

181

'Many times have I sat by the bedside of my patients for days together, giving it, anxiously waiting to see the effect it would have upon them,' he would write later in his book *The Theory and Treatment of Fevers*, while at the same time making a great show of administering a harmless placebo, a bread pill, for instance, to keep the confidence of both the patient and his relatives.

By the time Sappington left his childhood home in Tennessee to pursue his studies in Philadelphia in the second decade of the nineteenth century, he had become a firm convert to quinine's cause. He even persuaded one of the doctors at Philadelphia Medical College to try it. One afternoon over tea, his colleague Dr Barton mentioned that he had a most vexing patient. For three weeks the man had been in hospital suffering from 'simple ague'. He had been having fits of chills and fever, and although he was able to move about between the episodes, his condition wasn't really improving. Sappington said that he believed he could cure the man with one-ounce doses of bark in wine, 'as much as the stomach would bear, then doubling or tripling the dose at a time just before the return of the chill'. Barton followed Sappington's advice, and the patient was cured.

In 1823 a new company, Rosengarten & Sons in Philadelphia, began isolating quinine from cinchona bark on a commercial scale, following a method that had been perfected by two Parisian pharmacists, Joseph Pelletier and Joseph Caventou, three years earlier. What they called 'essence' of cinchona bark was in fact quinine sulphate, a soluble salt which had the advantage of being much easier to swallow than ground-up bark. Legend has it that when he heard of Rosengarten's new product, Dr Sappington immediately saddled his horse and rode all the way to Philadelphia. When he arrived, he sought out the company and bought its entire supply. After a few days, during which he attended some medical lectures, he returned home to Missouri carrying this wondrous substance in his saddlebags.

Whether this story is true or not is unknown. What is certain is that shortly after returning from Philadelphia, Sappington began using refined sulphate of quinine in his practice in addition to ground cinchona bark. He soon arrived at a curative dose of one or two grains of quinine sulphate every two hours for two or three days. For use as a prophylactic he prescribed the same dose, given four or five times daily.

In the spring of 1824, the efficacy of quinine sulphate was put to a severe test. On 16 May a large trading party, led by Meredith Miles Marmaduke, departed from Franklin, Missouri, bound for Santa Fe. Their first stop on the day after leaving Franklin was at the Sappington farmstead, at the start of the route known as the Santa Fe Trail, to purchase a supply of the 'essence of bark' that had been obtained in Philadelphia a few weeks earlier. There is a receipt for this purchase in the Sappington papers. In addition, the young Mr Marmaduke bade farewell to his fiancée, Lavinia Sappington, the doctor's daughter. Many years later, Marmaduke would become Governor of Missouri, but not before he had led the Marmaduke–Storrs trading party of eighty-three men safely across the malarious Missouri and Arkansas river valleys to Santa Fe, where they arrived without a single bout of illness to report, having carefully taken their daily dose of quinine as instructed by Dr Sappington.

The sick children in *The Little House on the Prairie* were treated by Mrs Scott, who 'poured a dreadful bitterness out of a small folded paper on to Laura's tongue'. Had they been travelling through Missouri in the 1830s, Mrs Scott would probably have dosed them with Dr Sappington's Fever Pills, the ingredients of which were a closely guarded secret, though the lid was clearly marked 'Price: One Dollar a Box'.

It is not clear exactly when Dr Sappington branched out from doctoring and farming and started his new business. By 1832, though, it was going strong. That year, according to the working formula sketched out on the back of a ledger book, he had perfected the secret recipe for his cure:

Sulphate of quinine 2 pounds
Pulv Extr. Liquerice 1 1/2 lb
Pulv Gum myrrh 1/2 lb
Oil of Sasafras
Acqua Pura
Make 240 boxes
24 pills to a box

Unlike the ancient Roman remedies that advocated incantations sung by virgins, or even Father Domenico Anda's brief mention of the Peruvian cortex in one of his concoctions at the Santo Spirito hospital, Dr Sappington's recipe was utterly practical. There was quinine as the active agent, liquorice to mask the bitter taste, sassafras oil for moistening the powdered ingredients, and gum of myrrh to bind it all together. In a long shed close to his farmstead, Sappington and, later, his sons and their slaves, would pound the mixture, press it into iron pill moulds, and pack the pills, twelve or twenty-four to a box, before stacking the boxes by the door ready for shipping.

Demand for the pills grew rapidly. By 1836, when Dr Sappington's son-in-law William Price travelled to Philadelphia to make his annual quinine purchases, he was ordering over 375 pounds of quinine sulphate at a time, more than a third of which had to be obtained in New York after the Philadelphia supply had been exhausted.

Sappington's sales force, particularly in the early years, consisted largely of family, friends and neighbours, all drafted in during the season to ensure the widest possible distribution. During the busier years, up to thirty-five people would set off, travelling for as much as six months of the year, to all parts of the central United States, from Ohio to Texas and from Nebraska to Alabama, distributing batches of pills to agents, who would sell them on for a commission. On major routes the agent could travel by steamer or by train, but once he penetrated deep into the less visited territories there was often no choice but to proceed by wagon or on horseback.

The letters the agents wrote back are full of annoyances and difficulties: extreme heat, flies and saddles that made the backs of the horses raw. At least two agents lost their horses altogether. A Mr Hall, having seen his horse killed by buffalo gnats, was forced to walk ninety-eight miles, at least part of it through a rising swamp. Augustus Stevenson was detained not only by rains and high water, but by the attempted theft of his bridle, martingale and saddle-blanket, which he retrieved by running after the thief and catching him. No agent provided such emphatic descriptions of his trials as George F. Bicknell, Sappington's agent in Mississippi, Louisiana and Texas. 'I have dragged like a snake my slow length along over hills and through mud to this place,' he said in one letter; and in another: 'I have been mightily persecuted with rain, roads impassible, bridges gone, creeks swimming . . .'. He was sure his horse would see 'affliction and infliction' before he was done with Texas, and ended his letter with the words: 'The Horse is alive but a little leg weary as is Yr Humblesvt.'

The profitability of the Sappington business relied on securing the main ingredient, quinine, at a reasonable price and on ensuring that the pills were properly paid for, which was not always easy in a new country where credit expanded and contracted, gold or silver were not always readily available, and paper money often fell in value. Despite their complaints, the agents pursued fresh sales with great success. In 1844, a good year, 500,000 boxes of fever pills were produced. Stories have been handed down that church bells were often tolled throughout the Mississippi valley at two-hourly intervals during the summer and autumn to remind all within hearing to take their anti-fever pills. Sappington, like Dona Phelicia Garzan with her shipment of bark from Peru to Spain more than a century earlier, had an effective product and a clientele that was willing to pay for it. As he himself would later recall in his book on treating fevers, he 'realised considerable sums of money' from the sales of Dr Sappington's Fever Pills.

For thirty years, the business thrived. But the coming of the Civil War would destroy Sappington's operation; quinine supplies became so intermittent that production virtually collapsed, as did his network of agents. Yet the authorities on both sides of the conflict, who saw their soldiers fall sick with malaria on almost a daily basis, would have benefited mightily had they paid the slightest attention to protecting and encouraging Sappington in what he was trying to achieve. In passing up this opportunity, they sentenced many of their soldiers to a lifetime of recurrent illness and, all too often, an early death.

Of all the grand civil engineering projects of the nineteenth century, none was more ambitious than the two canals dreamed up by the French at Suez and in Panama.

More than 5500 Frenchmen, along with another seventeen thousand workers from other countries, died in Panama between 1881 and 1889, during Ferdinand de Lesseps' ultimately ill-fated attempt to dig a canal across Central America. Because the Panamanians hold the French to blame for the fact that the canal belonged for nearly a hundred years to the United States of America, there is little to commemorate the work that de Lesseps and his compatriots were able to complete. Most of the bodies of the dead – other than those that were repatriated at great expense by their families – lie beneath the dark waters of the canal itself, along with more than seventy railway locomotives and thousands of tonnes of machinery and dredging equipment.

The main municipal cemetery of Panama City is in Chorillo, once one of the poorest quarters of the capital, where cinder-block public housing has now replaced the old wooden shacks of the canal workers. Within it rises one of the few remaining monuments to the period when the French were attempting to build the canal. The '*Recordación Perpetua a los Gloriosos Franceses Zapadores del Canal de Panamá*' records the names of twenty-seven men who died working for Ferdinand de Lesseps. They were born variously in Paris, Marseilles, Haute Savoie, Bresse, Tarn, Fontainebleau and, memorably, given the connection with de Lesseps, '*au barrage du Nil (Egypte)*'. They stand for thousands of others listed in the French records as '*décédé à Panama*'. Of the twenty-seven names, there is not one who was not a graduate of one of France's best engineering schools. Only two had reached the age of thirty. And they all died within weeks of arriving in Central America, in the winter of 1884–85, when the isthmus was swept by the worst epidemic it had ever known of yellow fever and malaria. Medicine was in short supply, and

thousands of workers died. The Compagnie Universelle du
Canal Interocéanique, for whom they worked, was so terrified
of what was happening that it forbade its employees from
disclosing the conditions under which they lived. One anxious
French shareholder who wrote to the company to ask about
the reports of excessive numbers of deaths, was soothed with
the words: 'We know you have confidence in us . . . If we say
there is no illness on the isthmus, this is because there is none,
and anyone who asserts the contrary is a gossipmonger who is
merely trying to undermine your confidence so that he can
take over himself.' Telling outright lies, the company decided,
was preferable to admitting the truth.

Philippe Bunau-Varilla, my great-grandfather, knew nothing about the epidemic when he sailed on the steamship
Washington from St Nazaire to the Panamanian port of Colón
in October 1884, although he made the crossing of the
Atlantic in the company of the Chief Engineer of the canal
project, Jules Dingler, and his family. It is likely that even if
Dingler, who had already served with distinction as the chief
of bridges and roads in France, had made a clean breast of the
problems they faced, the diminutive Bunau-Varilla would
not have listened. He had lived through the Franco–Prussian
war, and had seen people who were so hungry during the
siege of Paris of 1871 that they ate the animals in the zoo. He
was an outsider, a Protestant from the German borderlands
who had learned to make his way within the Catholic establishment, and he was so determined to break out of metropolitan France to make his fortune that he would have risked
any peril to achieve his aim. When asked by an indiscreet
company librarian why he should wish to commit suicide by
going to Panama, Bunau-Varilla had replied that he was
determined to go 'as an officer runs to [death] when he hastens

to the battlefield, and not as the coward who flees from the sorrows of life'.

Bunau-Varilla was born the son of a boarding-house owner from Alsace, and won a scholarship to the Ecole des Ponts et Chaussées, the engineering faculty of the Polytechnique de Paris, where he hid the fact that he was both a Protestant and illegitimate (*'de père inconnu'*, his records say) behind a sharply pointed moustache and a swagger that made one forget that he stood barely over five feet three inches tall. At the age of twenty-five he had achieved a good part of his ambition simply by getting to Central America, and nothing was going to stand in his way.

He had set his heart on the Panama Canal after hearing *le grand français*, Ferdinand de Lesseps, lecture at the Polytechnique. De Lesseps, who was still fêted throughout France for his triumph in completing the canal at Suez in 1869, appealed to the young engineers before him to pledge their lives to his grand new project for the sake of *la patrie et la gloire de la France*. Like many of those seated in the lecture hall, Bunau-Varilla needed no second call.

Had he been more sceptical, Bunau-Varilla might have asked de Lesseps some pertinent questions. Such as why, given that Panama was as rocky and mountainous as Suez had been sandy and flat, did he feel it was appropriate that the new canal should be built without locks? Why had no survey been carried out? What were the sanitary conditions of the isthmus? How were the sick to be cared for? What would happen if the costs ran over and the money ran out before the project was complete?

When the Dinglers disembarked at Colón in November 1884, with Bunau-Varilla close behind, it was raining so hard that they could not even see the hilltops that breasted the canopy of the jungle. To a first-time visitor, the mountains of the isthmus appear to be lined up in a continuous range, but as you approach them more closely you can see that they are really a succession of individual hills, arranged in bewildering complexity. Most mountain ranges were formed by the upward pressure of the earth's surface rising and folding. But the mountains of Central America are individual volcanic cores, little islands of hard rock such as basalt, diorite and andesite that pushed up through the layers of lava and ash. A cross-section of the isthmus shows not an orderly arrangement of layered ranges, but a profusion of faults, valleys, cores and dykes. In less than thirty miles, between Colón and Panama City, there are six major faults, five significant volcanic cores and seventeen different kinds of rock. It is not the easiest place to construct a canal.

In the course of the long windy days spent crossing the Atlantic, Dingler had told Bunau-Varilla a great deal more about the project, though he dwelt little on the difficulties that were growing worse by the day. One of these was the terrain they would have to work in; another was the climate. The rainy season came late in the winter of 1884, but it made up for the delay with a vicious intensity. I have never seen rain like the rain in Panama, not even as a child growing up in Africa, where we always believed we had real rain, rain that danced across the veld leaving a lemony light and a smell of new life. Even that was not the kind of rain that barrels down crumbling buildings, swelling the muddy torrents that swirl about so that you no longer can tell what is road and what is pavement. Panama rain is rain with *cojones*; what John le Carré calls 'six-inch raindrops, pumping up and down like bobbins on the front steps, the thunder and lightning setting off every car alarm in the street, and the drain covers bursting their housings and slithering like discuses down the road in the current'. I have never seen rain like the rain in Panama, and I bet my great-grandfather hadn't either.

'The points where my locomotive passed on the previous day were now covered by fourteen feet of water,' Bunau-Varilla wrote in his diary of a journey he made early on from the coast to Balboa in the centre of the isthmus, on the Chagres River. 'So I requisitioned three Indian canoes. We took these by train as far as we could, then embarked in them. As we paddled along through a channel apparently cut out of virgin forest all the workings were submerged and the tops of the telegraph poles were scarcely visible above the water.

'In places we had to drag the canoes over small hillocks, and in this way one canoe was damaged. We had then to crowd into the remaining boats, although the load was almost too much, and the freeboard was not more than an inch above water. One of the engineers, a M. Philippe, said that he couldn't swim. I told him jokingly, "There is no danger. I could easily tow you to the nearest trees." It was only then that I noticed the strangest phenomenon: the tops of the trees were not their usual green, but a distinct and ever-shifting black; as we drew nearer, I saw they were covered with the most enormous and deadly spiders: tarantulas.'

The flooding Chagres, whose level could rise fifty feet or more overnight, contributed not just to the workers' physical discomfort, but directly to the dangers in which they worked. De Lesseps' insistence that the canal would be built without locks had led the engineers to attempt to cut directly through the rock. At the highest point of the isthmus is the Gaillard Cut, then known as Culebra, a mile-long saddle of rocky land that soars three hundred feet above sea level. Its rocks consist of volcanic brescia, overlaid with a soft red clay that at first proved delightfully easy to dig away. 'He who did not see the Culebra Cut during the mighty work of excavation,' declared one witness, 'missed one of the great spectacles of the age – a sight that no other time or place was, or will be, given to man to see.'

In his arrogance, and because he had seen such an idea work on the terrain of the Corinth Canal in southern Greece, de Lesseps had decided that the walls of the Panama Canal would be cut through almost at the vertical. He had not reckoned on the rain that turned the clay to mud. Just one downpour was enough to send a slice of a newly exposed hillside slithering into the ditch that the French were trying to excavate. In mid-1883 the company was extracting the apparently impressive figure of more than 150,000 cubic metres a month – but de Lesseps had promised his investors that by that time the Compagnie would be carting away nearly ten times that volume.

By the end of 1884, when Bunau-Varilla got to work reorganising the dredging system on the deepest sections of the canal, matters had improved. But over the following year, more mud and rock fell into the ditch than it was possible to haul out of it. A total of 9.8 million cubic feet had been excavated by the end of 1884; after a further year of digging, dredging and carrying away of mud and rock, the total cumulative figure had been *reduced* by the end of 1885 to 9.4 million cubic feet. The French engineers were being assailed on every side. Machinery was constantly breaking down, and the sides of the canal would not stay in place. In the rainy season the rain never let up, not even for an hour. It forced up the level of the Chagres River, sent it flowing over its banks and cascading into the works that Bunau-Varilla and his companions thought they had completed.

With the rain came sickness. Never having been exposed

before to the mosquitoes that carried yellow fever or malaria, the workers on the Panama Canal were easy prey to disease. The workers refused to drink the rank water of the Chagres River, so the Compagnie carefully collected rainwater in empty tubs and bins at every street corner in the city. In the managers' gardens, precious shrubs and trees were protected from the tropic's voracious ants by tubs of water. And in the hospitals, black spiders were stopped from climbing into bed with the patients only when the bedlegs were placed in bowls of water. The rain filled ditches, holes and puddles. And wherever water lay undisturbed, the larvae of two kinds of mosquito were able to hatch: *Anopheles*, which carries *Plasmodium falciparum*, deadliest of the malaria parasites, and *Aedes aegypti*, or *Stegomyia fasciata* as it was then called, which transmits yellow fever.

Though both infections cause broadly similar symptoms – fever, headaches, vomiting – they follow different courses. Malaria, a parasitic invasion of the red blood cells, recurs constantly. Adults with malaria often survive, and if exposed to it over a long period, patients will build up some limited immunity. Yellow fever, by contrast, is caused by a virus that kills very fast. But patients who recover are then completely immune for the rest of their lives.

By December 1884, just a month after his arrival in Panama, Philippe Bunau-Varilla could see signs of illness all around him. 'Never had the yellow fever been more deadly,' he wrote in his journal. 'Some fathoms from my house at Colón, ships were anchored in the Company's harbour with not a single soul on board. All the crew had died.'

At first, it was inevitable that Bunau-Varilla, who counted himself blessed to be a Frenchman and who regarded survival as a sign of godliness, put a moral slant on the disease. 'The chief accountants of the two divisions placed under my orders were intimate friends,' he wrote.

They had come together to the Isthmus, they lived in the same house and ate at the same table. One of them was an irreproachable character, while suspicions clouded the other's reputation. Public rumour, when I arrived, imputed to him certain reprehensible acts. They concerned the payment for some heads of cattle bought for feeding our workmen, when the interruption of the train service by the insurrection nearly caused a famine in Colón. People spoke of fraudulent payment to the purveyors with the complicity of the accountant. A rapid inquiry strengthened the accusation in my mind. I ordered the incriminated man to appear before me on the following morning at seven o'clock. His no doubt heavily burdened conscience, as well as the certainty of finding no mercy from me for a breach of honour, so troubled him that he fell ill with fever and was taken to the hospital.

Eight days later he was carried to the cemetery. His comrade and friend, who was unaware of the cause of his disease, was terribly frightened when he left him to go to the hospital. 'If he dies,' said he, 'it will be my turn next.' But in the event he did not succumb.

Many a time I went to see the ships as they arrived from Europe filled with employees. Many a man on them had been happy to enlist, but felt his heart sink at the sight of the warm, low and misty shores of the deadly Isthmus. Some bore on their faces the obvious mark of terror. I often took note of their names to see how they would stand the trial. Without exception they were dead within three months, if they had not fled the Isthmus.

For every eighty employees who survived six months on the Isthmus one could say that twenty died.

Bunau-Varilla's observations were harsh, and his conclusions hardly scientific. By the end of 1884, more than a hundred workers were dying each month in the two main hospitals. This was ten times the death rate two years earlier, but given that many more men died at home or on the construction site, the real figure was much higher. William Gorgas, the army doctor who would take over running the medical programme while the canal was being built by the Americans in the early twentieth century, reckoned that for every death recorded in French hospitals there were at least two more outside that were not counted. De Lesseps, *le grand français*, began to be called 'The Great Undertaker'.

The months after their arrival were terrible for the Dingler family who crossed the Atlantic with Bunau-Varilla. Their daughter was the first to succumb, to yellow fever. A dark-haired, pretty girl of eighteen, she liked riding side-saddle into the hills wearing a full skirt and a little Panama hat. 'My poor husband is in a despair which is painful to see,' Madame Dingler wrote to de Lesseps' son Charles. 'My first desire was to flee as fast as possible and carry far from this murderous country those who are left to me. But my husband is a man of duty and tries to make me understand that his honour is to the trust that you have placed in him and that he cannot fail in this task without failing himself. Our dear daughter was our pride and joy.' Jules Dingler, his twenty-one-year-old son and his daughter's fiancé rode at the head of a long procession to the cemetery.

A month later, Dingler's son showed signs of the disease. Within three days, he too was dead. Presently, the fiancé died, also of yellow fever. And on 1 January 1885, at the height of the epidemic, Madame Dingler too succumbed. Ever the public servant, the Chief Engineer was at his desk by seven the next morning. But Dingler's grief finally caused him to lose his mind, and he sailed back to France. In the absence of anyone else available, Bunau-Varilla was appointed in his place.

Within a few months, though, it was Bunau-Varilla's turn to fall ill. At the end of March, he lunched with a group of French officials from the Quai d'Orsay who had come to view the progress of the canal. Arrangements were made to meet the following morning to begin the tour of inspection. That night, Bunau-Varilla's valet, George Octave, was awakened by the sound of his master tossing feverishly in bed. Octave ran to the nearest cottage, where some of the canal's technicians slept, yelling: 'The *calentura*! Master Philippe has the *calentura*.' Within half an hour, a doctor and a retinue of attendants had arrived.

Bunau-Varilla had seen from his hospital visits how few patients recovered from the dreaded yellow fever or malaria. To prevent the former he swallowed a concoction given to him by a local *curandero*, or healer, while for the latter he took regular doses of quinine. If he fell ill, he instructed his servants, under no circumstances was he to be moved from his quarters. On the third day of his confinement in his room he was visited by a French doctor. Sitting up, Bunau-Varilla observed: 'I am doing as well as a man can do when approaching the cemetery.'

'Why do you say that, *Monsieur le Directeur*?' asked the doctor.

'Because at the start, my pulse was sixty to the minute, the following day it had fallen to fifty, yesterday to forty . . . and today it is thirty. In three more days I will not be here to make any more observations about myself.' Knowing that Bunau-Varilla was in the habit of taking quinine and the treatment prescribed by the *curandero*, the doctor suggested the dosages of both be increased. Before long, the patient fell asleep, and by the following morning the decline in his pulse rate had been reversed. Within days, Bunau-Varilla was up and about.

My great-grandfather's self-prescribing, though it may have worked, casts little light on the practice of medicine in nineteenth-century Panama. The French established two main hospitals, one in Ancón, close to Panama City, and the other on Limón Bay, which surrounds Colón, the port on the Atlantic side where Bunau-Varilla arrived in Panama. Other than that there were small medical outposts and dressing stations at various posts along the railway line that crossed the isthmus. Sponging and bloodletting were still the favoured treatments for fever even as late as the 1880s, though there were people who swore by bourbon and mustard seed as an 'infallible specific' for yellow fever. Fever was regarded as an ailment in its own right, caused by an imbalance of the humours, rather than as a symptom of disease. Malaria, or intermittent fever as it was called, was said to be caused by the miasma. Yellow fever, or the *vómito negro*, for its black vomit, was considered to be directly contagious, and anyone diagnosed with it could be certain to have his bedding and belongings burned within a few hours. The true nature of parasitic disease and the role of the mosquito in transmitting both malaria and yellow fever were completely unknown, and the insects were allowed to flourish unchecked. There were no screens at the windows, and no attempt was made to isolate patients with contagious diseases – hospital patients were allotted beds by nationality rather than by illness.

De Lesseps' Compagnie did not place a premium on

keeping its workers healthy, and although the price of quinine had begun to slide a little after the American Civil War, it still traded at nearly five hundred French gold francs a kilo. The management believed it was cheaper to replace sick or dying workers with fresh men than to spend a lot of time and money treating them with quinine in hospital. It thus instituted a policy of ignoring quinine altogether, and refusing to dose its workers. The only quinine available was what the engineers, usually the most senior among them, had brought with them for personal use. My great-grandfather's supplies were sent to him regularly from Paris by his mother.

In 1889 Ferdinand de Lesseps' Panama Canal project collapsed, and the Compagnie Universelle du Canal Interocéanique was declared bankrupt. The last straw was a bond issue that failed. On the morning of the subscription a bogus report swept the globe announcing that de Lesseps had died in his sleep. De Lesseps himself describes its effect: 'Lying rumours and false telegrams announcing my death were circulated all over the world! The scoundrels had chosen their time well; for a denial could not be issued until too late . . . A complaint has been lodged with the public prosecutor, who has started an investigation into these criminal acts.' The culprits were never found, and de Lesseps' last-ditch attempt to save his company, a whirlwind tour of France to drum up support for the bond issue, ended in an impassioned plea: 'I appeal to all Frenchmen. I appeal to all my associates whose fortunes are threatened. I have dedicated my life to two great works which have been called impossible: Suez and Panama. Suez is completed and France has been enriched. Do you wish to complete Panama? Your fate is in your hands. Decide!' Only half the bonds were taken up. The public had decided.

By the time the Americans took over the building of the

Panama Canal in 1903, much had changed. Panama itself, which had formerly been only a province of Colombia, had seceded. Its revolution was protected by American warships ordered there by Theodore Roosevelt, for Washington had been offered the canal and a strip of land on either side, the canal zone, in return for its backing. The Americans had been prompted to become involved after one of their ships, the *Maine*, was blown up in Havana harbour at the start of the Spanish–American war in 1898. The only ship that could come to its rescue, the *Oregon*, had been forced to sail twelve thousand miles around Cape Horn for lack of a Central American canal. The journey took sixty-seven days, and the *Oregon* arrived as hostilities were coming to an end.

'If we are to hold our own in the struggle for naval and commercial supremacy,' Roosevelt later told a Chicago business club, 'we must build up our power without our borders. We must build the Isthmian canal, and we must grasp the points of vantage which will enable us to have our say in deciding the destiny of the oceans of the east and west.' His bravado notwithstanding, Roosevelt was denounced in the press for not having consulted Congress more before supporting Panama's secession. He turned to one of his more humorous cabinet colleagues for support:

'Have I defended myself?'

'You certainly have, Mr President,' came the reply. 'You have shown that you were accused of seduction and you have conclusively proved that you were guilty of rape.'

Philippe Bunau-Varilla set out early on to enlist Roosevelt's help. He saw off a rival claim to build another canal across the isthmus in Nicaragua, by sending a postcard with a Nicaraguan stamp to every member of Congress and anyone else he could think of besides. The stamp showed the Nicaraguan volcano Mount Momotombo in full eruption. On the card he wrote, 'An official witness of the volcanic activity on the isthmus of Nicaragua.' If there was one thing a trans-isthmian canal did not need it was a live volcano nearby; any hint of one was bound to frighten off potential investors.

In November 1903 a treaty formalising the annexation of the canal zone was drawn up in Washington. John Hay, the American Secretary of State, signed it on behalf of the United States, Bunau-Varilla for the French. In a letter to his eleven-year-old daughter, my grandmother, he described his life's work as 'the struggle I kept up for the defence and triumph of the greatest moral interest which the French genius has ever had abroad'.

French genius there may have been, but the US Army applied different methods to making the Panama Canal project succeed. Where there was chaos, it instituted order; it upgraded the technology, rationalised the machinery used on site and carefully marshalled thousands of workers. But nowhere did the army make more difference than in the methods it deployed for fighting disease.

This was no straightforward or easy task. Colonel George W. Goethals, who took charge of the canal works after Panama's secession, was aghast at the cost of maintaining the health of his workers. Much of the sanitary work being done was on the basis of the newly propounded theory that malaria and yellow fever were spread by mosquitoes.

'Do you know,' Goethals once asked the head of the sanitary department, William Gorgas, 'that every mosquito you kill costs the United States government ten dollars?'

'But just think,' Gorgas replied. 'One of those ten-dollar mosquitoes might bite you, and what a loss that would be to the country.'

Despite Goethals' efforts at thwarting him, Gorgas had the support of President Roosevelt, who had appointed him with the instruction that he should do whatever was necessary to wipe the isthmus clean of disease.

William Crawford Gorgas was born in Mobile, Alabama, in 1854, the son of a Confederate Chief of Ordnance. He followed his father into the army, and throughout his life he remained a doctor first and an army officer second. During the Spanish–American war he served in Cuba, initially assisting on the experimental work which established the connection between fever and mosquitoes, then on the constantly developing precautions for eliminating malaria and yellow fever. In Havana he worked closely with Carlos Finlay, a doctor of Franco-Scottish extraction living in Cuba, who as early as 1881 first put forward the theory of a link between yellow fever and the mosquito. Gorgas carried out a number of experiments in an effort to turn Finlay's inspired guess into proven fact. He had volunteers bitten by mosquitoes, injected with parasites and toxins, and made to sleep in vomit-covered bedding surrounded by bowls of excreta donated by yellow-fever patients. Some of these selfless volunteers contracted yellow fever; some even died. But their sacrifice was not in vain. Early in 1901 the Army Medical Board was able to state categorically that yellow fever was transmitted by the *Aedes* mosquito, that this mosquito must previously have fed on the blood of an individual sick with fever, that an interval of at least twelve days after feeding on the blood of a yellow-fever victim was needed before the mosquito could pass on the disease, that the period of incubation after a bite was between forty-one and 137 hours, and that infection could not be passed on by clothing, bedding or vomit.

In March 1904, Gorgas was appointed Chief Sanitary

Officer to the Isthmian Commission, at the instigation of President Roosevelt. He was based at the hospital in Ancón, from where he was determined 'to do what I could to prevent the enormous loss of life which had attended the efforts of the French'.

Gorgas was initially hampered in his work by bureaucracy, lack of funds and deep scepticism about the mosquito theory. 'The whole idea of mosquitoes,' the chairman of the Isthmian Commission told him, 'is the veriest balderdash.' The Governor of Panama was, if anything, even more direct: 'I'm trying to set you straight, Gorgas. On the mosquito you are simply wild. Get the idea out of your head. Everyone knows that yellow fever is caused by filth.'

It was not until 1905, nearly two years after the Americans took over the canal, that Gorgas was given the go-ahead he needed to carry out Roosevelt's instructions. What had changed the climate of opinion was yet another outbreak of yellow fever in the winter of 1904. Although it killed no more than eighty-four people, the epidemic showed how very easily the American-run canal zone could be threatened by the same disease that had seen off the French. A new Chief Engineer, John Stevens, was appointed, and on 2 August, just a week and a day after his arrival, he told Gorgas that he could have all the money, supplies, equipment and men that he needed. Ten years later, after the canal was opened, Gorgas, by then the US Surgeon-General, wrote privately to Stevens:

I have a very clear and grateful recollection of the support and friendship you always gave me on the isthmus. I knew very well that you were the only one of the chief officials who believed in the sanitary work we were doing, and who was not taking active measures to oppose us. The fact is that you are the only one of the higher officials on the isthmus who always supported the Sanitary Department, and I mean this to apply to the whole ten years, both before and after your time, so you can understand that our relations, yours and mine, stand out in my memory of the very trying ten years I spent on the isthmus as a green and pleasant oasis.

The first task Gorgas set himself was to eliminate malaria and yellow fever from the isthmus. He used many of the methods he and Carlos Finlay had worked on in Havana. In the late summer of 1905 he attacked the breeding grounds of the mosquito. The use of water containers within a hundred yards of any piped supply was prohibited. Ponds, cisterns and cesspools were to be coated with oil once a week. Where water barrels had to be used, they must be screened. It was made a punishable offence to harbour mosquito larvae, even if done inadvertently. Inspectors made house-to-house searches for the larvae, and a poor report would quickly be followed by a visit from commission carpenters, who repaired sagging gutters, put up screens of copper-wire mesh, and emptied out infected *tinafas*, stoneware jugs, and the bowls of water that helped protect hospital patients from enthusiastic spiders.

The inhabitants of Panama City submitted with good-humoured, if bewildered, resignation, though they were less amenable to the second plank of Gorgas's campaign: fumigation. Initially, only those houses in which a case of yellow fever had been diagnosed were fumigated. Gorgas knew from first-hand experience that a mosquito had to have fed on the blood of a yellow-fever carrier to become a carrier itself. But it was only when it was seen how effectively fumigation destroyed mosquitoes and hampered further breeding that fumigation was extended to all houses in Panama City and Colón. By the summer of 1906 every building in both cities had been meticulously sterilised.

In the first twelve months of Gorgas's sanitary rampage, his squads got through a thousand tons of timber and two hundred tons of copper-wire mesh. They burned three hundred tons of sulphur and 120 tons of natural insecticide made from pyrethrum – the entire output of the United States – in an effort to smoke themselves into cleanliness. The results were worth it. In 1905 there were more than two hundred cases of yellow fever in Panama; the following year Gorgas's team counted only one. Not a single case was registered from 1907 on.

Gorgas attacked malaria in much the same way as he dealt with yellow fever. He eliminated the breeding ground of the *Anopheles* mosquito, destroyed their larvae and killed the full-grown mosquitoes. He had the labour force sleep under mosquito nets in rooms whose windows were protected with narrow wire meshing. His final weapon was quinine. Shortly after he arrived, he persuaded the Isthmian Commission to overrule the French ordnance that quinine was too expensive to be given to canal workers; henceforth all the workforce would be protected by quinine.

In the course of the winter of 1905, Gorgas ordered 2874 pounds of quinine sulphate, enough to treat every worker in the canal zone. He had it administered as a prophylactic, in doses of 0.15 grams twice a day. The white Americans working on the canal were quick to appreciate how well quinine kept malaria at bay and how effective it was in lessening the severity of attacks. Among the Caribbean workers it was far less popular, until Gorgas had the idea of dissolving it

Sir Clements Markham, President of the Royal Geographical Society, whose exploration of the Peruvian Andes would lead him to try to establish cinchona plantations in India's Nilgiri hills (*right*).

Richard Spruce, the Yorkshire plant collector whose exploration
of the Orinoco and the Amazon would add greatly to our
knowledge of plant life in South America.

(*Above*) Charles Ledger, an East End chancer, whose smuggled *Cinchona ledgeriana* seeds would grow into the richest quinine-bearing trees in the world, and (*right*) Manuel Incra Mamani, the Bolivian guide who first found the hoard and whose quest to find more ended in tragedy.

The Missouri doctor, John Sappington, whose 'Sappington Fever Pills', made of quinine, would travel in the saddlebags of Meriwether Lewis and William Clark on their journey to open up the American West in 1803.

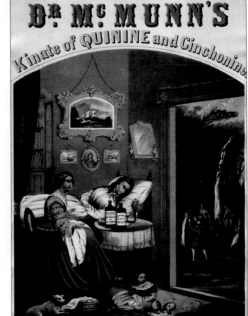

A woman's work is never done. A nineteenth-century advertisement for quinine, the nurse's friend.

in extra-sweet pink lemonade. In making the drug more palatable, he did exactly what the officers of the Indian army had done in creating quinine tonic water half a century earlier, when they added sugar to the bitter quinine they were made to swallow to prevent malaria, then added a little gin to the mixture: the original gin and tonic. In the second half of the nineteenth century, the descendants of a German jeweller based in Geneva, Johann Jakob Schweppe, used their newly patented bubbling device on a mixture of oranges, sugar and quinine. They called it Schweppes Indian Tonic Water.

It is hard to judge the efficacy of quinine alone in Panama. But Gorgas's sanitary campaign, combined with his prescription of quinine as a preventive, reduced the incidence of malaria by a considerable degree. In 1906, the year he began his campaign in earnest, 82 per cent of the workforce was found to harbour malaria parasites in their blood. When the canal was completed in 1913, fewer than 8 per cent of workers showed any sign of malaria at all.

Although no one still knew exactly how the disease was transmitted or how quinine cured it, as at Walcheren, in the Confederacy and on the Niger, the Jesuits' powder was an essential armament for those engaged in both exploration and warfare. Without it, the expansion of the British Empire, the control of the seas, the extension of trade, war and colonial adventure into the tropics would have followed a very different, and perhaps narrower, path. But quinine was still difficult to obtain, and expensive when it was available. The question of how to ensure a cheap and regular supply was vexing colonial administrators and bounty hunters from as far afield as Chile, Goa and Kew Gardens in London.

The Seed – South America

'Cinchona succirubra *is a very handsome tree, and
looking out over the forest I could never see any other tree
at all comparable to it for beauty.'*

RICHARD SPRUCE, botanist, 1860

*'There is no drug more valuable to man than the febrifuge
alkaloid which is extracted from the chinchona tree of
South America.'*

SIR CLEMENTS MARKHAM, KCB, FRS,
*Peruvian Bark: A Popular Account of the
Introduction of Cinchona Cultivation into
British India* (1880)

At the time that doctors, merchants and politicians finally
became convinced of the crucial role quinine had to play in
treating malaria, and therefore in helping military and ex-
ploratory expansion overseas, there were still no plants to be
found outside the forests of the Andes, nor even any seeds.
That was about to change.

In his memoir about Peru, to which he travelled after being
commissioned to find the rich red trees of the *Cinchona
succirubra* and bring seeds and saplings back to Britain, Sir
Clements Markham, later President of the Royal Geographical Society, describes the soft-spoken Yorkshire plant hunter

Richard Spruce. He was, he writes, 'an eminent botanist and most intrepid explorer . . . I shall ever look upon my good fortune in securing Dr Spruce's able co-operation as the most fortunate event connected with my conduct of the enterprise.' The two men shared a passion for plants and plant hunting, especially in South America, and for the cinchona tree in particular. Although they worked closely together for only three years, they came to owe each other a great deal, which is why, decades later, Markham intervened with some of his influential political friends in London to secure a pension for the then impoverished and sickly Dr Spruce. Notwithstanding all his efforts, Spruce's pension turned out to be almost derisory, but in the last years of his life it was his only income.

Despite their shared Yorkshire heritage, their common interests and the successes they enjoyed together, the two men could not have been more different in character. Clements Markham was a Victorian metropolitan – politically wily and socially ambitious. He demonstrated his achievements in a particularly English way, collecting so many honours to follow his name that, after his knighthood, his Fellowship of the Royal Society and the accolades from Portugal, Brazil, Sweden and Norway, the remainder were briskly dealt with by the mock-humble appendage, 'etc., etc.'.

Richard Spruce too could not have been anything but English. A tall north countryman who played the bagpipes, the soft-spoken bachelor was as renowned for his modesty as for his encyclopaedic knowledge of mosses. A photographic portrait taken on his forty-seventh birthday shows a neat man with long legs and diffident lips hidden by a beard that had not yet grown as long as it would in later years. Yet, like so many Englishmen who only truly become themselves when released from the strictures of home, there was another side to

Spruce. Finding himself far up the Amazon one night at a small village that was preparing to celebrate the Feast of St John, Spruce was invited to dance by a local dignitary. She happened to be a woman. Later he recalled that when he saw that 'it was intended to do me honour and that I should be accounted proud if I refused, I led the lady out, first casting off my shoes in order to be on terms of equality with the rest of the performers. We got through the dance triumphantly, and at its close there was a general *viva* and clapping of hands for the "good white man who did not despise other people's customs!" Once in for it, I danced all night.'

Spruce had an innate interest in other people, whatever their customs, but only in South America could he cast off his natural English reserve as easily as he had his shoes on that hot Amazon night. His warmth drew others to him. Twelve years after Spruce first visited the Ecuadorean town of Ambato, between Quito and Guayaquil, his former landlord, Manuel Santander, wrote to him: 'Come to your Ambato to lay your bones along with ours . . . Oh if we had you at our side, we should be happy.'

Richard Spruce first set foot in South America in 1849. He had to wait until he was thirty-two to make the journey, for he was poor, and uncertain whether he could earn his living as a botanist. His interest in the Amazon had been awakened ten years earlier by reading Charles Darwin's account of his voyage on the *Beagle*, but his fascination with plants had started when he was a boy. The son of a Yorkshire schoolmaster from Ganthorpe, Spruce lost his mother when he was still young. When Spruce was fourteen, his father married again, and had eight daughters, which robbed the boy of all but the most marginal parental attention. Unable to do much for his only son other than suggest that he follow in his footsteps and become a schoolteacher, the elder Spruce never-theless encouraged his early interest in plants, and even accompanied him on long walks across the Yorkshire moors.

By the time he was sixteen, Spruce had drawn up a list, neatly written out in alphabetical order, of 403 plants he had gathered and named from around Ganthorpe. In 1841 he dis-covered and identified as a new British plant a very rare sedge, *Carex paradoxa*. By then he had also begun a study of native mosses. A year earlier, the first edition of Henry Baines's *Flora of Yorkshire* had been published, which included no more than four mosses. Yet in a single three-week excursion to Teesdale, the sharp-eyed Spruce identified another 163 mosses and forty-one hepaticae, of which six of the mosses and one *Jungermannia* were new to Britain.

Before he left for South America, Spruce spent a year on the border between France and Spain, searching for plants in the Pyrenees. Contemplating his future on his return to London, he was confident he would never have to go back to schoolteaching. The Pyrenean expedition had been the suggestion of the English botanist George Bentham and his friend Sir William Hooker, Director of the Royal Botanic Gardens at Kew, who had learned of Spruce's existence after his success in discovering the sedge *Carex paradoxa*.

The growing passion of the age for science, for cabinets of curiosities and for collecting in general, meant there was an increasing number of wealthy patrons prepared to pay such a fine botanist as Spruce for dried plants that were well preserved and accurately named. Hooker was right to be con-fident of Spruce's talent. The young Yorkshireman had taken with him to Spain a guide book which stated categorically that '*la famille de mousses n'existe pas dans les Pyrénées*'. But within a short time Spruce had gathered together 230 hitherto unknown mosses and three hundred species of other choice alpine plants for despatching to purchasers in Britain and on the continent.

It was Bentham and Hooker, whom he visited on his return to London in 1846, who suggested that the greatest territory still awaiting a skilled naturalist collector such as Spruce was the Amazon. That, the two men insisted, was where he should go next. Within a few months Spruce set sail for Pará, on the coast of Brazil. When he returned to Britain a decade and a half later in 1864, he had collected more than thirty thousand plants.

The Royal Botanic Gardens at Kew, which Sir William Hooker had schemed for so long to take over and which he ruled with a firm hand and unbridled ambition, had been transformed during the first half of the nineteenth century. It had already grown enormously by 1800 – in 1768, when James Cook's *Endeavour* sailed for the South Seas, there were six hundred species of plants in cultivation at Kew; thirty years later the figure had grown to more than six thousand – but at the turn of the century it was still little more than a royal pleasure ground, a botanical sandpit where a gentleman amateur, in this case King George III, could indulge his private passion for natural philosophy and show off to his friends.

Kew was in many ways similar to that other royal garden that had been created in the middle of the Prado by the Spanish King, Charles III, in Madrid in 1774. If there was any real public function at all to Kew, it was as a symbol of the power and modernity of King George; the garden reflected the dignity of the monarchy, and its vast collection of orchids, palms and rare trees added lustre to the court. 'When their Majesties resided at Kew,' wrote a contemporary historian, 'a Terrace near the river was frequented . . . with a concourse of nobility and gentry! Stars and ribbons and garters glistened on the eye in uninterrupted succession.'

After the death of George III in 1820, Kew suffered a brief eclipse. The new King, the urbane George IV, could not abide the thought of mud, and swiftly began building himself a new home, Buckingham Palace, in town. Before long, though, Kew was once again a centre of ambition and ambitious expansion. Britain had already set about a regular annexation of foreign parts – territories in the West Indies and the Orient were followed by Aden in 1839, and New Zealand, Hong Kong, Natal, Punjab and Basutoland in subsequent years. It was natural that a policy of exchanging plants with these newly acquired territories would soon take hold. So too did the tradition of financing botanists to travel abroad. When William Balfour Baikie travelled up the Niger aboard the *Pleiad*, it was in the company of a former gardener from Kew, Charles Barter.

By 1840, the tide of economic and political reform that was seeping into every corner of English life had pervaded even Kew's elegant promenades. That year, the ownership of the gardens was transferred from the direct control of Queen Victoria to Her Majesty's government. Although no longer owned by the monarch, Kew, and by extension British botany, still enjoyed its royal patronage. One of the earliest acts of the Botanical Society of London, when it was founded in 1836, was to present Victoria with her first banana.

How Kew flourished under Hooker, its first botanical President, and later his son Joseph, is a parable of the politics of reform and the expansion of empire in nineteenth-century Britain. Central to the garden's reinvented self was its first imperial project, the British cinchona initiative.

In 1858 the share announcement of the newly formed Central Africa Company described in some detail how the loss of European life that had always accompanied any foray into Central Africa had deterred many individuals from opening up trade there. That problem, the announcement promised, had now been overcome. Henceforth it would be possible to expand African commerce, to offer scientists and missionaries a wider field for their operations, and thus eventually to improve 'the moral and material condition of the [African] people'. What made all this possible, the document stated, was the efficacy of quinine in preventing intermittent fever.

The idea that cinchona might be grown in other European colonies had been suggested as early as the eighteenth century. The Spanish, naturally, were vehemently against such a plan, for it would mean the end of their monopoly. In 1835 it had been proposed that the mountainous parts of India, in particular the Khasia and Nilgiri Hills, might be the place to attempt cinchona cultivation. Nothing more was done at the time, but during the winter of 1852–53 a Dutch botanist, Justus Karl Hasskarl, who had at one time been the superintendent of the Buitenzorg garden in Java, travelled to South America disguised as a German businessman, with the intention of collecting cinchona seeds for planting in Java. Although Hasskarl arrived too late for the flowering season and was forced to wait in Peru for another year, his presence in South America accelerated a race in which Britain, Holland and France sought to secure the best varieties of cinchona for plantations in Asia.

Early in 1858, five years after Hasskarl's visit and shortly after the share announcement of the Central Africa Company, British businessmen in Ecuador enlisted the help of the local *chargé d'affaires* to prod the government on the matter. In lieu of more than £1 million in overdue debt and interest payments, the government of Ecuador had granted to English bondholders 4½ million acres of land, which constituted the major asset of the Ecuador Land Company. In a memorial to the Board of Trade in London, the businessmen stated that 'the transportation of the cinchona tree to India and some other British colonies . . . forms part of our contemplated undertaking'. The new India Office, which had been created in the wake of the Crown's assumption of control over the subcontinent after the Indian Mutiny of 1857, offered an enthusiastic endorsement of the scheme.

Kew was to play a major part in the cinchona initiative, handling the administration of all stages of the project, the systematic collection of the best varieties of cinchona in South America, the germination of seeds in British greenhouses, and the transportation of the plants and their introduction into botanical gardens in British Asia.

Also crucial to the project were Richard Spruce, who was already collecting plants in South America for his private subscribers, and the as-yet unknighted Clements Markham, who was then employed as a junior clerk at the Board of Control, copying out letters and despatches from India, Persia, Syria and the Orient, but who earlier, in the course of a ten-month visit to Peru in 1852–53, had travelled extensively through the country. Markham was the first Englishman to visit Cuzco, and had taught himself Spanish and even a little Quechua grammar, but his most important claim was that he was one of the few Europeans to have seen a cinchona tree, *Cinchona ovata*, the species that the Dutch botanist Hasskarl would later try to smuggle to Java, thus starting the race to secure the tree's best varieties.

It is likely that Markham, ensconced within the bowels of the Civil Service, with its informal network of spies and bartering of influence, would have heard whispers about the cinchona initiative early on. The Board of Control worked closely with the English East India Company, the powerful enterprise that had come to control most of India by the early nineteenth century, and that had long been an enthusiastic supporter of the idea of growing cinchona on the subcontinent. In 1852, at the same time that Markham was travelling through Peru, and flushed with the success of his Indian pavilion at the Great Exhibition, John Forbes Royle, the Reporter on Indian Products to the East India Company, wrote to his superiors warning them that the *Cinchona calisaya* forests in South America were being cut down so fast they would soon be destroyed. In March 1857, after Baikie had proved quinine's virtues as a prophylactic against malaria while sailing up the Niger, Royle again wrote formally to the Company: 'The almost inappreciable value of the cinchona, commonly called Peruvian bark . . . is universally acknowledged. Hence it becomes a duty to humanity . . . to increase the supply of cinchona trees which yield such valuable barks.'

The Board of Control's ties with the India Office meant that Clements Markham would have been well aware of the problems that endemic malaria, unchecked and seemingly uncheckable, was causing in India. Even though he was neither a doctor nor a botanist, he would have understood immediately the full implications of Royle's memorandum, which seemed on the face of it to be simply philanthropic. But to anyone with the slightest acquaintance with the problem, Royle's words had a commercial and political undertone that was all too clear. To ensure the health of Indian workers, a reliable source of quinine, locally and cheaply produced, and easy to administer, was needed. Plants and seeds would be required to start off a plantation, and these would have to come from South America.

Although he was only twenty-eight, Markham was a Fellow of the Royal Geographical Society, thanks to his friend, the Society's President Sir Roderick Murchison. His account of his Peruvian trip, *Cuzco and Lima*, had been widely reviewed when it was published in early 1856. Ambitious, well connected and supremely self-confident, Markham began an energetic letter-writing campaign in support of botany for Britain – with the aim of having himself appointed to lead the expedition to South America to secure the best varieties of cinchona.

Many years later, Markham would be captivated by the story of the Countess of Chinchón. Unwilling, perhaps out of snobbishness, to accept that it might not be true, he did much to promote the tale in the English-speaking world. In 1886 he even went so far as to try to persuade the International Botanical Congress to change Linnaeus' botanical nomenclature of the tree from the erroneous *Cinchona* to *Chinchona*.

It is characteristic of Markham that he should be attracted to a story that had a countess as its heroine. His fondness for titles, and the natural ease he displayed when cultivating men of influence, were the result of an early unconscious training. Three years before Markham was born in Yorkshire in 1830, his father, a clergyman, had been appointed canon at Windsor Castle. With the job came a house within the castle walls, and by the time Clements was a young boy his father was dining every Sunday with the King and Queen, wearing evening dress with knee-breeches, silk stockings and shoes with silver buckles. Canon Markham would bring back to his family bonbons from the royal table, and an endless supply of gossip.

At the age of twelve, Markham was sent to Westminster School, the *alma mater* of many earlier Markhams. He was already showing an interest in geography, devouring accounts of polar voyages, and also history. By that time he had written a history of England, geographies of various countries including Peru, studies of astronomy and a number of short biographies. At the same time he met Lord Ellesmere, like Murchison – and indeed Markham himself – a future President of the Royal Geographical Society. Both Ellesmere and the masters at Westminster encouraged Markham to write further.

Soon after, Markham was introduced to another man who would further his career. At a dinner given by his aunt, the Countess of Mansfield, he met Rear-Admiral Sir George Seymour, then a Lord of the Admiralty, who suggested that Markham enter the Royal Navy and accompany him on his flagship to the Pacific. Less than a month later, having taken an exam that consisted merely of writing down the Lord's Prayer, Markham was enrolled as a naval cadet and assigned to HMS *Collingwood*. On 20 July 1844, Markham's fourteenth birthday, the *Collingwood* sailed out of Portsmouth harbour.

Over the next six years, Markham would travel as far afield as the Falklands, Chile, Peru, Tahiti, Hawaii, and even the Arctic. He read and reread accounts of the voyages of William Dampier, Captain James Cook, and W.H. Prescott's *Conquest of Peru* (1847). Despite the excitement he felt at each new stop, he grew increasingly disenchanted with what he saw as the random and excessive cruelty of the Navy. In 1851, after several attempts, he finally persuaded his father to allow him to resign his commission. But it was not until he returned to Britain that he told his family of his ambition to return to the place that had so captured his imagination when he first docked there as a young naval cadet in January 1845: Peru.

On 20 August 1852, a month after his twenty-second birthday, Markham set out once again from Windsor. He never saw his father again; he would learn of his death the following year when he returned to Lima after his journey through central Peru and read his father's obituary in an old copy of *The Times*. Having left home, Markham proceeded to Liverpool, and from there crossed to Halifax with two former messmates from the *Collingwood*. When he shook hands with them at Windsor, Nova Scotia, 'and really started on my Peruvian expedition', he wrote later, 'I felt that then I finally left the navy.'

Markham's principal aim during his travels in Peru was to learn as much as possible about the Incas. He had already developed a number of theories about their origins and culture, which he wanted to test by visiting, measuring, drawing and describing the ancient ruins that dotted the Peruvian landscape. He was also fascinated by the Spanish conquest, in particular by the Inca chronicler Garcilaso de la Vega (1539–1616), whose writings he had pored over even though he had only a second-rate translation. But it was impossible for him not to be moved by the dynamic optimism with which Peru, and in particular Lima, had been seized in the years following its independence from Spain in 1825.

Markham's itinerary would take him across the isthmus of Panama, where the railway that preceded French efforts at dredging a canal was still being built. Where there was no line, a road – 'the most execrable in the world' – provided the only way ahead. The climate, as attractive to the mosquito as it was unpalatable to human beings, was as bad as ever. 'We were overtaken by heavy showers,' Markham wrote in his journal, 'the rain coming down in buckets, clothes being converted into a soaking sponge, and so we went on – heavily, despondently.'

Life improved when Markham docked at Callao. He found Lima little changed since he had last seen it in 1847, although he noted that the *say y manto* dresses – with their narrow pleated skirts and their odd hood arrangement that covered the head and face, leaving only one eye showing – were no longer to be seen, having fallen from fashion in the meantime. He dined, thanks to numerous introductions from London, with the cream of Limena society, and attended a ball in the finest house in the city. Finally he set off on horseback on his journey. Asked by a passing colonel, as he left town, whether he was going to *pasear* as far as Magdalena, three miles distant, Markham replied, 'No! To Cuzco,' which was three hundred miles away. '*Caramba*,' replied the colonel, clearly impressed. '*Dios guarda vos.*'

Markham's travelling kit closely resembled that which Agustino Salumbrino and the other Jesuit fathers had used when setting off from San Pablo a century and a half earlier. He also took with him a sextant, hygrometer, telescope and, thanks to an Englishwoman who was concerned about his welfare, three pots of sardines.

The influence of the Jesuits was still clearly visible to Markham, though they had been banished from the country eighty-five years before. As he left the coast and began to climb towards the *cordillera* and Cuzco, Markham passed well-tended estates, vineyards, sugar mills and handsome churches, all of which had been built by the Jesuits. By the time he reached the Peruvian *montana*, the area on the eastern slopes of the Andes where the cinchona grows, he was well acquainted with Peru's agricultural economy. There he saw trees of India rubber, vanilla, indigo, balsam, cinnamon and sarsaparilla; indeed, many of the products that formed the bulk of the cargo Dona Pheliziana Barbara Garzan had watched being loaded aboard the *Conde* at Guayaquil in 1748.

It was in the *montana*, where the spurs of the Andes reach down into the vast tropical forest of the Amazonian basin below, that Markham caught sight of his first cinchona tree one day in early May 1853: 'Gradually the slopes covered with long grass were exchanged for a subtropical vegetation. There were many beautiful flowering plants. I here saw a cascarilla tree and afterwards another. The species was *Cinchona ovata*, not a valuable kind, but it made me acquainted with the genus.'

Five years later, when, back in England, he began campaigning to be named as the head of the cinchona expedition to South America, Markham wrote to the Under-Secretary of State for India. Although he was no botanist, he described at some length how he had learned Spanish and Quechua, and how well he knew the bark forests of Peru and the landowners of the eastern *cordillera*. He also argued that it was essential for the leader of such an expedition to be a man 'whose heart [was] really in the business'. The more he thought about it, the more fired up Markham became about the possibilities that John Forbes Royle's proposal to his superiors at the East India Company had suggested. 'I gave the subject most careful consideration,' he wrote later, 'and being convinced that this measure would confer an inestimable benefit on British India, and on the world generally, I resolved to undertake its execution.'

Markham's own suggestion was that the India Office should finance four separate expeditions to gather the eight most valuable varieties of cinchona growing in the Andes. In addition to both Bolivian *calisayas* and the two varieties of red bark growing in Ecuador, he proposed that plant collectors should be sent to Peru and to Colombia. 'If the thing is worth doing,' he wrote to the India Office, 'it is worth doing well.'

In the end, three expeditions only were approved: Markham's own, a second one to Huánuco, also in Peru, and a third to Ecuador. This last, it was decided, should be led by the moss man, Richard Spruce.

While Clements Markham was climbing through the ranks of the British Civil Service, trading a titbit of information here or extracting a favour there, Richard Spruce was focusing all his attention on the miraculous abundance of the Amazon. It seemed that everywhere he turned he spotted not just his beloved mosses, but new flowers, canes, grasses, ferns, trees, lichen and palms, not to mention animals and birds, many of which had never previously been recorded. 'Here our grasses are bamboos sixty or more feet high,' he wrote to his family. 'Our milkworts are stout woody twiners, ascending to the highest tops of trees . . . Instead of your periwinkles, we have here handsome trees exuding a most deadly poison. Violets are the size of apple trees . . . Daisies borne on trees like Alders.'

Spruce's first introduction to the economic potential of cinchona came from a druggist named Daniel Hanbury, whose family ran a thriving pharmaceutical business in Spitalfields, in the East End of London. In 1856, Spruce received a letter of introduction from Hanbury. It had been forwarded to South America by the botanist George Bentham, who described Hanbury as 'a young man of great abilities who has devoted himself much to pharmacology and is anxious on the question of cinchona etc upon which he writes to you'. Hanbury urged Spruce to collect young cinchona plants and also the 'old bark', that was rich in alkaloids: 'Botanical specimens of Cinchona & its allies, accompanied by small, authenticated specimens of bark, could not fail to meet with some demand here and on the continent in quarters where the general herbarium would not be subscribed.'

Three years later, when Spruce received the stiff envelope containing the commission from the India Office, he had been in South America for nearly ten years. His health, never good, caused him much trouble. Furthermore, he was broke, for he never charged his subscribers overmuch for the specimens he sent them. In March 1857 he had set out on a trek, in the midst of revolution, from the town of Tarapoto, in northern Peru, along the steep banks of two rivers that swirled and fell in muddy cascades down the western slope of the Andes, into Ecuador. This ghastly journey lasted 102 days, leaving him shrunken and emaciated from lack of food. Before that, in 1853, he had tramped through the floodwaters at the edge of the Amazon, where bloodsucking bats – 'midnight bloodletters', he called them – left large patches of dried-up blood, drawn from previous travellers, on the floor of his room. He had turned north at the rubber boom-town of Manaus and canoed over much of the Río Negro, through the country explored by the German botanist Baron von Humboldt fifty years before, and as far as the cataracts of the Orinoco River in southern Venezuela. The spray of the waterfall 'dashes in one's face, and [the] roar drowns one's voice', he wrote. Mosquitoes, known locally as *zancudos*, were a constant scourge for both Spruce and his guides. 'If I passed my hand across my face,' he wrote from the town of Esmeraldas, at the foot of Mount Duida, to a friend back in England, 'I brought it away covered with blood and with the crushed bodies of gorged mosquitoes.'

At Maipures, further up the Orinoco, the mosquitoes were so bad that Spruce was forced to wear gloves and to tie his trousers down over his ankles. 'I constantly returned from my walks with my hands, feet, neck, and face covered with blood, and I found I could nowhere escape these pests.' Even eating in that place was no pleasure. 'Many times there is no sitting down to eat a meal, but one must walk about, platter in hand, and be content to eat one's food well peppered with mosquitoes.'

Like everyone else at the time, Spruce made no connection between mosquitoes and malaria. For someone who complained so much about his health, he travelled with remarkably little in the way of medicine. Yet, if he hadn't had a small bottle of quinine with him on that trip up the Orinoco, he would surely have died. After only a few days at Maipures, Spruce began to feel unwell: 'I had violent attacks of fever by night, with short respites in the middle of the day, and on the second night, on stepping out of my hammock, I was seized with vomiting . . . I passed a dreadful night, and in the morning I resolved to seek better aid.' A local *comisario* and his nurse advised certain local pills, one a violent purgative, which would surely have exacerbated Spruce's symptoms. He refused to touch his small store of quinine – perhaps he did not really believe he had malaria, and was afraid to waste what little quinine he had with him. In any event, when his condition showed no signs of improving after two weeks, he changed his mind.

At first he took only two or three grains. As his symptoms improved, he increased the dose, until by the nineteenth day he was taking six grains at least four times a day. The effect was, as it almost always is at high dosage, virtually instantaneous. His fever fell and his appetite returned. Within two days he was able to sit at table; within a week he was dining on tapir. Soon after, he asked a visiting manufacturer of feather hammocks he knew, Antonio Diaz, to help him get back downriver to San Carlos. 'A few weeks ago I thought I would never write to you more,' Spruce later confessed to Bentham. 'In seeking the wherewithal to sustain life, I came within little of meeting with death.'

Spruce's encounter with malaria left him not just physically weak, but poor as well. Six years later this helped seal his determination to take on the India Office commission, which he did swiftly, asking just £30 a month – £10 more than he usually received from his subscribers – as a salary. Although he would in later years come to regret having accepted so little, Spruce then had eyes only on the immediate future. 'I have been entrusted by the India Government,' he wrote from his new base at Ambato in eastern Ecuador, a short distance south of Mount Cotopaxi, to his colleague John Teasdale in April 1859, 'with the charge of obtaining seeds and young plants of the different sorts of cinchona (Peruvian Bark) found in the Quitonian Andes for transporting to our Eastern possessions, where it is proposed to form plantations of these precious trees on a large scale. This task will occupy me (if my life be spared) the greater part of next year.'

Having recovered from the perilous journey along the eastern Andes to Ambato, Spruce was ready to seek the cinchona. He hired an Indian guide named Bermeo, 'who had worked a good deal at getting out *cascarilla*', and set out with mules and horses to scout about. Although he found much else to interest him, including a lush oasis of vegetable-ivory palms and a small tree called *Palo del Rosario* which he said had never been described before, Spruce was disturbed by the extent to which the cinchona trees had been destroyed. He found several lying on the ground, and the only one still standing had been stripped of its bark 'near the root, so that it was dead and leafless'. He did not give up, though. Through the good offices of James Taylor, an English doctor he met in Ambato, and funds put forward by the British Consul, Spruce was after much negotiation able to lease a large acreage of forest belonging to a former President of Ecuador, General Juan José Flores, a patient of Dr Taylor's. For a rental of $400, Spruce would be allowed to take as many cinchona seeds and plants as he liked, on condition that he did not touch any of the bark.

He secured the services of four Indians who were used to working in the bark forests, and began loading up provisions of potatoes, peas and barley meal, as well as the boxes of precious botanical equipment. The expedition used the tough Andean bulls called *cabrestillos*, whose cloven hoofs made them more secure than mules on the slippery inclines they would have to traverse while passing beneath the summit of Mount Chimborazo. Leaving winter behind on the eastern side of the Andes, Spruce and his guides headed westwards across a saddle towards Guaranda. The vegetation – hassocks of *stipa* and *festuca* – gave the landscape a grey, barren air. But there were hundreds of other plants: lupins, calceolarias, coloured violets and geraniums, as well as brightly coloured green shoots that grew merrily out of the carcases of old, cut-down trees. These were cinchonas that had been harvested by earlier *cascarilleros*, but which were being lovingly tended by the Indians who now knew their true worth.

Having reached the small hamlet of Limón, where the finest clutch of red-bark cinchonas was once known to have grown, Spruce first stopped to kill one of the bulls so as to have a supply of dried beef, then set about visiting all the bark trees in the vicinity. He was glad to see that many had reached their full size, and that most of these boasted a thick crop of flowers and the beginnings of the seed capsules he was hoping to empty. As June turned to July, the weather became cooler, with 'a good deal of mist and fog'. The capsules were attacked by maggots, and many of them started to turn mouldy. 'I began to fear we should get no ripe seeds,' Spruce wrote later in his journal. Things grew worse when he found one morning that two of the best trees had been stripped completely bare by the people of Limón, who had heard that he wanted to buy seeds. 'I immediately went round to the inhabitants and informed them that the seeds would be of no value to me unless I gathered them myself.'

Spruce found about two hundred trees in the valleys around Limón. Only two or three of these were saplings, the remainder being regrowth from old trees that had been cut down earlier. He was unable to find a single young plant under these trees. Many seeds germinated in cane fields, where they failed to survive the frequent weeding, or in pastures where they were grazed by cattle. The bark hunters or *cascarilleros*, Spruce learned, rarely went out before the month of August because of the fog. Once out on the slopes, though, they worked fast, cutting down trees and digging out many of the roots in order to strip off the bark with a machete, in much the same way that oak used to be stripped in England. They then built a stage about three feet high, known as a *tendál*, for drying the bark, taking special care that neither rain nor the fire below the *tendál* posed any danger. Once the bark was dry they carried it to a nearby depot, where they were paid $20 per quintal, about one hundred pounds.

Although the *cascarilleros* knew that the bark was worth money, they had no idea what it was used for. Most of them believed it was transported to foreign countries where it was somehow made into dye that was either, depending on who you asked, the colour of coffee or, even more valuable to the Ecuadoreans, the dark sheen of chocolate. 'I explained to the people of Limón how it yielded the precious quinine which was of such vast use in medicine,' Spruce wrote in his journal, 'but I afterwards heard them saying to one another, "It is all very fine for him to stuff us with such a tale; of course *he* won't tell us how the dye is made, or we should use it ourselves for our ponchos and *bayetas*, and not let foreigners take away so much of it." '

Despite their scepticism, the Indians were keen to help. In late July, Spruce learned with some relief that the Wardian cases, the heavy wood and glass boxes in which he planned to transport the young cinchona plants, were finally on their way. Their safe passage was being overseen by Robert Cross, a young gardener whom the Royal Botanic Gardens at Kew had sent to South America. He arrived 'pale and thin', not to mention anxious, for there was much unrest in the towns, and he had passed many troops along the way.

Cross set to work taking cuttings of the red bark cinchona. The owner of a nearby *chacra*, a small farm, had shown Spruce some sprigs that he had cut from the stool of an old tree and stuck into the ground by a nearby watercourse four months earlier, and 'they had all rooted well'. On 13 August Spruce noticed that some of the finest seed capsules were beginning to burst. An Indian climbed up the tree, and breaking the panicles gently, let them fall on sheets that had been spread on the ground. Cross sowed eight of these seeds, and he and Spruce watched over their hatchlings with all the excitement of parents awaiting the birth of a first child.

One began to germinate on the fourth day, and at the end of a fortnight four had put forth their first leaves. In time, they all followed. 'One of the seedlings was afterwards lost by accident,' wrote Spruce, 'but the remaining seven formed healthy little plants, and when embarked at Guayaquil, along with the rooted cuttings and layers, bid as fair as any of the latter to reach India alive.'

In all, Spruce and Cross collected more than 100,000 ripe seeds, and more than six hundred cuttings and seedlings. The seeds they dried in the sun before carefully packing them in boxes with dried leaves. Cross took charge of packing up the seedlings, taking care to wrap each one in wet moss before placing it carefully with the others in a basket that was balanced, one on each side, of a bull. 'There [were] not a few falls on the way, and some of the baskets got partially crushed by the wilfulness of the bulls in running through the bush; but the greater part of the plants turned out wonderfully fresh.'

For all the thousands of miles that the seedlings were to cover on their way first to Jamaica and then to England, before departing again for India, it was the early part of their journey by river to Guayaquil that was to prove the most perilous. Intermittent heavy rain and flooding higher up the mountains sent the level of the river shooting up and down 'so that we had to watch our raft night and day', wrote Spruce. With the help of three raftsmen wielding heavy wooden oars they swept along, their raft bobbing over the river surface like a dancer. In six hours they had covered nearly twenty-five miles, when the river suddenly narrowed to less than thirty yards, and the raft with its precious cargo was thrust into a boiling sluice that threatened to submerge it.

Fearful that the Wardian cases might be smashed, Spruce and Cross elected not to put on the glass covers, but to stretch strips of moistened calico over the top of the frame instead. 'The heavy cases were hoisted up and dashed against each other, the roof of our cabin smashed in, and the old pilot was for some moments so completely involved in the branches and the wreck of the roof, that I expected nothing but that he had been carried away.' The cases received only a few slight cracks, and none of them had turned over, wrote Spruce. But his greatest worry, as ever, was for his small charges. 'The leaves of the plants were sorely maltreated,' he wrote, but they survived happily enough. By the time they were embarked aboard a large steamer at Guayaquil, Spruce's only concern was that the warm sea atmosphere was making them grow too rapidly.

While Spruce's quest, difficult though it had been, was ultimately successful, Clements Markham's journey across Peru would have quite a different outcome.

Markham set sail from Southampton on 17 December 1859. Though the weather was grey and windy, the damp sky with its fishy south coast smell was as familiar to the small party as the South American landscape would later prove exotic and frightening. Travelling with Markham were his young wife Minna, a clever, spirited young woman whose sense of adventure and curiosity easily matched her husband's, and John Weir, the second of the two gardeners seconded by the Royal Botanic Gardens at Kew to ensure the cinchona plants survived in good health. Christmas was spent in the middle of the Atlantic, and by January the growing numbers of brown pelicans whirling about in the air told them they were approaching land. Sometimes by mule and sometimes by rail, Markham's party crossed the isthmus of Panama, and then, like Robert Cross less than a year earlier, caught the steamer south towards Lima.

If Markham, who was neither a botanist nor a plantsman

like Weir, felt nervous about the task ahead, he showed no sign of it. His old friends in Lima welcomed him as if he had never been away, and a local paper, *Commercio de Lima*, published a glowing review of his book about Cuzco and the Incas. Between parties and receptions, Markham made himself busy arranging for supplies and transport, and after a month the expedition was ready to begin.

The party sailed south to Islay on the Pacific coast, then headed inland towards Arequipa. It was the same route Markham had taken seven years earlier, and he still had friends in the pretty hillside town, where he arranged for his wife to stay while he and Weir travelled on. So intent was Markham on forging ahead that he gave little thought to how circumstances, both Peru's and his own, had changed since his last visit. The young Englishman whose heart had soared at the thought of escaping the suffocating embrace of the Royal Navy, and who had travelled to Peru to study and to learn, was not the man who was returning now. Markham in 1860 came with all the financial and political backing of the Victorian establishment, and much of its self-righteousness as well. 'In obtaining plants and seeds of the valuable chinchonas from South America, it would have been a source of deep regret to me if that measure had been attended by any injury to the people or the commerce of Peru, Ecuador, or Colombia,' he would write later. 'But,' he went on – adopting the line of thought that justified so much that the Victorians did, from saving the Elgin marbles to forcing Africans to wear European clothes – 'hitherto they have destroyed the chinchona trees in a spirit of reckless short-sightedness, and thus done more injury to their own interests than could possibly have arisen from any commercial competition.'

The Peruvians might well have destroyed much of their

cinchona forests by cutting down the trees willy-nilly, but they also held their miraculous fever-trees in high regard: when the country became independent in 1825, its national emblem boasted a small woolly vicuna, a horn of plenty and a cinchona tree. By clothing his mission in a mantle of philanthropy, Markham blinded himself to the notion that others might have a different perception of his efforts.

As Markham left Arequipa and began the climb through the clouds of the *cordillera* on the western side of the Andes, he had other things on his mind. Both he and Weir had begun suffering terribly from the *sorochi*, or mountain sickness, that travellers to Cuzco are still warned about today: 'It began with a violent pressure on the head, accompanied by acute pain and aches in the back of the neck, causing great pain and discomfort, and these symptoms increased in intensity during the night at the Apo post-house, so that at 3 a.m., when we recommenced our journey, I was unable to mount my mule without assistance.'

Despite the discomfort, Markham and Weir pressed on, drawn as much by the novelties in the landscape as by the challenge they had set themselves. As they crossed stream after stream of freezing clear water, they spied herds of vicuna grazing on the slopes, or galloping along at great speed with their noses close to the ground, 'as if scenting out the best pastures'. They saw huge numbers of plovers, uttering their discordant notes as they flew overhead before turning to skim the ground in circles. Green parrots chattered unseen in the trees, with partridges and brightly coloured finches. But what really caught their imaginations was the glorious *coraquenque*, the condor, the 'royal bird of the Incas', with its striking black and white wings that had been so vividly described by the Inca chronicler Markham admired so much, Garcilaso de la Vega.

Markham's rapture was interrupted only by the mules, which were a constant source of trouble. He had refused to engage an experienced muleteer, thinking to economise by managing the animals himself. He soon found out his mistake. 'Whenever the brutes had a chance,' he told his biographer, 'they would bolt off the road in different directions, bumping their packs against the rocks or endeavouring to roll, which, of course, would soon have smashed everything they were carrying.'

Markham had hoped that by striking eastwards from the sea he would eventually make his way towards La Paz in Bolivia, where the best *Cinchona calisaya* trees were known to grow. But by the time he and his party were approaching the Bolivian border, they had more than mountain sickness and mad mules to cope with. Near Crucero, in the western *cordillera*, they met a red-faced and choleric old man, a former colonel in the Peruvian army named Don Manuel Martel, who had begun spreading the rumour that the English travellers were preparing to steal the Peruvians' precious bark. Martel, it seemed, had lost a great deal of money in the quinine trade, and he had nothing good to say about Justus Karl Hasskarl, the Dutch agent who had come to Peru in disguise six years earlier in an attempt to smuggle some seedlings to plant in Java. If the Dutchman, or anyone else, ever attempted to take the *cascarilla* out of the country again, Martel threatened, he would stir up the people to seize them and cut off their feet.

Finding the border closed as a result of one of the many intermittent disputes between Peru and Bolivia, Markham resolved to restrict his hunt for *Cinchona calisaya* to the western side of the border. From the mountain town of Sandia, his party headed down into the Tambopata valley. By night, it drizzled without stopping. There was little dry ground on which to pitch a tent, and almost nowhere to light a fire for a cup of tea or beef concentrate. By day, they had to force their way through the forest, slashing through closely matted ferns, fallen bamboos and the roots of enormous trees. Small black flies and other insects stung the exposed skin of their wrists and ears. The men used their machetes to slice at the creepers whose long tendrils twisted around their ankles, threatening to trip them up at every step. After half an hour of hacking and chopping, they would often look back to find they had advanced no more than twenty yards. The sweat ran down their necks, and the sticky yellow mud clung to their boots. In many places the forest was so overgrown, and the foliage so dense, that it was almost dark even at noon, except where a few gaps allowed the sun to shed a pale light across their gloomy surroundings.

One day followed another as Markham, Weir and their Indian guides searched for cinchona plants. The Indians rolled coca leaves into balls and chewed them while they worked. Markham adopted the same habit, finding that the soothing effect enabled him to climb the steep slopes and to cope more easily without food. For more than a fortnight they searched the cinchona forest, and by early May they had gathered four round bundles of cinchona plants, five hundred in all. It was enough to fill the Wardian cases that were waiting to be despatched to the coast.

As Markham and his party climbed out of the forest on 12 May, they were stopped by a gang of young men who had been lying in wait for them. Leading them was the son of Don Manuel Martel. The young man shouted at Markham's Indian guides, accusing them of betraying their nation by helping to convey the plants out of the country. As soon as they reached Sandia, he threatened, the plants would be seized and confiscated. Markham raised his revolver, though he knew that, like everything else he possessed, after spending a fortnight tramping through the Andean undergrowth it was so damp as to be useless. The young man opposite him did not know that, though. Markham looked him in the eye, and let out a steady breath. After a long moment, Martel's son stepped aside to let Markham and his men pass by.

Knowing that it was only a matter of time before he would be stopped again, Markham contemplated setting out on foot alone, with the four bundles of plants on his mule. But when he reached Sandia he was approached by a man whose acquaintance he had made on his first trip to the town. If Markham would hand over his gun, the man told him, he would supply him with an Indian guide who could find him fresh beasts and accompany him to Vilque, on the road to Arequipa.

Markham readily consented to this plan, and despatched Weir with another guide in the opposite direction to throw Martel off the scent. On 17 May he left Sandia with his guide, Angelino Paco, and two mules bearing the precious plants. But it became clear before night fell that Paco had never been away from the valley of Sandia, and had no idea how to find his way across the *cordillera*. Moreover, when they stopped and opened their bags, Markham found that all his food and his matches had been stolen. That night he shared Paco's dry maize meal.

The two men resumed their march at daybreak. For nearly twelve hours they wandered across the grass-covered plains of the high *cordillera*. At nightfall, they found a deserted shepherd's hut built of loose stones and thatched with wild grass. It stood no more than three feet high, but both men flung themselves inside it to shelter from the wind and the cold.

Next morning they found that the mules had wandered off, and it took them three hours to find them and round them up. The mules required constant supervision. If left to themselves, Markham wrote, they would try to lie down and roll over, which would have been 'fatal to the plants that were strapped on their backs'.

For more than two weeks Markham shepherded his saplings towards the sea, stopping briefly at Vilque for fresh supplies, and at Arequipa, where he was reunited with the patient Minna. But the party's arrival at Islay, on the coast, brought little respite. The customs master declared that it was illegal to export the *cascarilla* plants, and refused to allow them to be shipped without an express order from the Minister of Finance and Commerce. Markham left the saplings behind and caught the first steamer to Lima, hoping that the Ministry would know little about Peru's ever-shifting rules about trade. Only a few days earlier the President had issued a decree forbidding the export of cinchona, but luckily for Markham, the Minister did not yet know this. After much cajoling, Markham obtained the papers he needed, and promptly left once again for Islay, where he found his plants happily budding and throwing out leaves inside their Wardian cocoons, proof that they had survived their journey across the Andes.

For all his efforts, Markham's expedition was not a success. Of the 529 plants he shipped from the forest of Caravaya, nearly half died before he and Weir reached Southampton in August 1860. Instead of taking those that remained to Kew, where they could safely recuperate and grow stronger, Markham insisted that they proceed straight to Bombay and the Nilgiri Hills in south-western India, where they were due to be planted out in the newly established cinchona plantation. It was a terrible mistake. Some of the cases were knocked about in the crossing, and one was dropped into the Red Sea. It was by now late summer, and a hot wind was blowing eastwards from the Sahara. At one point the ship's engines stalled for more than a day, and the temperature on deck rose higher and higher. On reaching Bombay, Markham discovered that the connecting steamer he had been due to board for Calicut had already left, and he and his precious cargo were left to wait in the heat before being able to continue their onward journey.

It was not until mid-October that the plants finally reached the government plantations near Mysore. There, with the help of a skilled plantsman, William McIvor, who was the superintendent of the botanical gardens, they were planted out in fresh soil. But it was too late. By December, exactly a year after Markham had sailed to South America from Southampton, every one of them was dead.

A copy of a portrait of Sir Clements Markham by George Henry hangs in the Royal Geographical Society. It was commissioned by the governors of Westminster School when Markham was eighty-three, and at his suggestion it includes an illustration of *Cinchona officinalis*, the highly prized *cascarilla fina*, and a silver model of Captain Scott pulling his sledge towards the South Pole, the two geographical exploits that made Markham's reputation. Markham would write at length about his journey to find the cinchonas of South America – his *Peruvian Bark*, which was published in 1880, runs to more than five hundred pages. Yet neither there, nor in any of the articles he wrote, nor even in his private journals and the two volumes of cinchona notebooks that are lodged with the Royal Geographical Society, does he mention the sad end of his *calisaya* seedlings.

One reason may be that he learned that Spruce's plants were the only ones that had been successfully transplanted in India. More galling, though, was the fact that the seedlings that had been collected by both Spruce and Markham were almost valueless. Rival plantations started by the Dutch in Java would be filled with cinchona trees grown from seeds procured by another Englishman, Charles Ledger, and his Bolivian guide, Manuel Incra Mamani.

Charles Ledger was a short, barrel-chested man with a re-markable tenacity that served him well in his years as a frontiersman in South America. Like so many Europeans who travelled to South America in the nineteenth and early twentieth centuries, he relished the freedom offered by a society that observed few rules. To survive he had to be imaginative, resilient and immensely resourceful. Yet he was capable also of harbouring deep resentment, and nothing troubled him more about an enterprise than the sense that while the effort had all been his, the reward went elsewhere. When in later life, having fallen on hard times, he wrote to friends in London asking them to exert their influence to secure him a pension, his letters reveal that for him there was no slight too small to be ignored, no grievance too old not to be recalled. Yet the part he played in obtaining cinchona and introducing it to the world belies any suspicion that he was filled with deep wells of self-pity.

Like Markham, Ledger was not a trained plant hunter. But by the time he entered the hunt for the cinchona tree he was more adept at the task than Markham would ever be. Born in Bermondsey in south London in 1818, Ledger called himself 'a true cockney'. His family were traders and merchants working on the fringes of the City of London. He had enough schooling to read and write, but his agility with figures he would pick up on the job. On his eighteenth birthday in March 1836 he sailed for South America, working his passage across the Atlantic and carrying in his pocket two letters of introduction and the fruits of a £10 note that had been given to him by a family friend who happened also to be the Mayor of London. Young Charles had spent his gift on a consignment of steel pens, then quite a novelty, which he would sell for a profit when he arrived in Buenos Aires.

After more than eight months at sea, Ledger docked at Callao, in northern Peru, the same port that had welcomed the Jesuit pharmacist Agustino Salumbrino almost two and a half centuries earlier. From there he proceeded not to either of the companies to which he would be introduced by the letters he bore in his pocket, but to a third firm he thought he would like more, an enterprising merchant house called Naylor Kendall & Co. Ledger was taken on right away, and put to work as a clerk. He had been teaching himself Spanish out of books while on board ship; now he was made to master book-keeping, administration and everything there was to know about alpaca wool and cinchona bark, the two products that Naylor's traded most.

Ledger quickly completed a rudimentary apprenticeship, upon which he was put to work in the warehouse where consignments of merchandise were received from the interior. He learned how to grade cinchona bark and to sort wool according to its quality and colour. And he had to ensure, when the bark was packed up to be sent to Liverpool, that it was properly bundled into the *serones*, fresh bullocks' hide sewn with thongs, that would keep it safe from rain and mildew on its journey to Europe, while at the same time never letting it dry out so much that it crumbled into dust.

Soon Ledger was allowed to travel into the interior. He was sent to Naylor's second office in Tacna, not far from the southern border with Chile and just down the coast from Islay, where more than two decades later Markham would load up the *calisaya* plants he had found in Caravaya province in the western *cordillera*. Before long Ledger was scouting the countryside, making contacts and learning the ways of the Indians who sold him their wool and their bark. He was not slow to realise that while he had to put up with poor food and uncomfortable lodgings, or no lodgings at all, in his efforts to secure fresh produce, most of the profit from his trading went to Naylor's. Upon becoming engaged to the daughter of a prominent Tacna merchant, he decided to leave the firm, and once married he set up on his own. Naylor's recognised in Ledger that same spirit of enterprise and independence on which the firm itself had been founded, and its owners were sensible enough to encourage his ambitions rather than try and stop him. Soon he had become one of their biggest suppliers, seeking out new Indian contacts in the rarely-visited areas along the border with Bolivia, who would bring him their bark and wool.

One of the Indians Ledger liked most was Manuel Incra Mamani, a Bolivian *cascarillero* he met in La Paz in 1841, the year before he left Naylor's to set up on his own. Mamani was a small, quiet man, but the readiness with which he was prepared to leave his native valley to work for Ledger made him unusual. The more Ledger came to know him, the more he realised that Mamani made up for his lack of words with an impish sense of humour and a deep knowledge of the forests and valleys in which they travelled. At a glance Mamani could distinguish between different cinchonas that would appear virtually identical to anyone else; on one trip alone he pointed out twenty-nine different varieties – Hugh Algernon Weddell, in his classic book on South American flora *Histoire naturelle des quinquinas*, published in 1849, described only nineteen. For hours on end, walking alongside Ledger's mule, chewing coca while Ledger smoked his pipe, Mamani would identify the differences between varieties of cinchona, depending on the shape, size and colour of their leaves, which changed with the season and the age of the tree, and the colour and shape of the flowers when the trees bloomed in December at the beginning of summer. Occasionally they would come across a group of five cinchonas planted in the shape of a cross, and Mamani would fall on his knees, cross himself and offer a prayer to *los buenos padres*, the Jesuit fathers who so long ago had sought to teach the *cascarilleros* to replace each tree they cut down with five more.

The *cascarilleros* had long since forgotten the priests' message. The forests were being stripped of their cinchona trees, and no one was giving any thought to ensuring future supplies of bark. As a result, sources of good-quality cinchona were becoming harder and harder to find. Yet demand for the bark continued to increase, as did the price. Two years after Ledger set up as a trader on his own, the Bolivian government issued an edict restricting the cinchona trade to three merchants who together paid fifty thousand silver Peruvian dollars to secure a monopoly for three years. The agreement doubtless enriched the government officials who had signed it; it also gave birth to a resourceful and imaginative trade in smuggled *cascarilla*. Ledger, whose reputation in the area was growing, and who hoped to identify out-of-the-way virgin places where the best trees could still be found in numbers, was one of the first to become involved. In 1845, a syndicate of Peruvian merchants clubbed together, each contributing five hundred Peruvian dollars to a common fund and providing a servant with a saddle and pack mule to take part in the first expedition. There were two guides. With Mamani as his companion, Ledger struck north from Puno in the company of fifty-five other hopeful smugglers, and headed for the Huánuco valley.

Despite their efforts, the merchants failed to find a single tree that produced the true *calisaya* bark. 'After 57 days of excitement, dangers, toil and disappointment', Ledger wrote, they returned to Puno. Yet something had happened in the course of that journey that would give Ledger much to think about.

One night while they were camping high on the eastern *cordillera*, it was Ledger and Mamani's turn to take guard. The summer had not yet begun, and it was still quite cool. Mamani sat chewing the inevitable coca leaves, while Ledger drew on his pipe. For a while they watched in silence as the mist slowly drew up to cover the forest below them. Above, the night sky was clear, and the moon bathed the forest and the rocks in a lemony-coloured mist.

'Will we find the true bark, do you think?' asked Ledger.

'No, *señor*,' Mamani answered firmly. 'The trees that grow here do not see the snowy caps of the *cordillera*.'

Ledger dismissed Mamani's comment. 'I could hardly contain my laughter at that moment,' he would later recall.

The two men bade each other goodnight. Ledger lay on the low camp-bed that Mamani had prepared for him, and pulled his poncho around his shoulders. The moon cast dark shadows upon the trees around him. High above, stars pricked the sky. As the fire died down, Ledger found himself thinking again about what Mamani had said. Mamani could spot cinchona trees just by looking at the forest canopy. He was able to recognise the true *calisaya* simply by its foliage, when everyone else found one *calisaya* indistinguishable from another. Mamani could tell before chewing it which variety a fragment of bark was. He must have had a reason for what he said.

The next morning, Ledger unfolded the maps he always carried with him and traced the spine of the Andes, and the *cordillera* that spread like ribs on either side from Quito to Nueva Grande in Colombia, and to Loxa in Ecuador. This was the cinchona's homeland, along with the valley of Caravaya in Peru where Markham would find his plants, and further south-east into Bolivia. The best red bark came from the regions that were dominated by mountain peaks: Illimani and Sorata, both in western Bolivia, and the massive Chimborazo volcano that Spruce would explore. Around Puno, where Ledger and his syndicate were travelling, the peaks were lower, and as Mamani had said, there was no true *Cinchona calisaya* to be found.

Dejected by the failure of the syndicate's expedition, Ledger turned his hand to trying to make his fortune out of that other South American staple, alpaca, the soft, downy wool of the vicuna. He journeyed through the countryside buying animals here and there, slowly making his way southward towards Chile. He travelled on foot, chivvying his beasts through the high mountain passes, gathering them close around him when the snow proved too deep to pass through, and turning them out into grassy pens whenever he reached spring grazing. He even made preparations to ship his alpaca to Australia, where he thought he might start a new life. Yet the thought of the true *calisaya* and the curling white petals of the cinchona tree never completely left him. While exploring a possible route for the export of alpaca in the southern summer of 1851, shortly before the young Clements Markham would see his first cinchona in a valley east of Cuzco, Ledger and Mamani were visited by the spectre of the miraculous fever-tree.

The two men had set off from Puno, just as they had done with the syndicate six years earlier. From there they crossed into the Bolivian Amazon through a region that was so densely forested they had to carve their way through the roots and vines. Not even a mule could find its way through the undergrowth, and Ledger and Mamani were forced to proceed on foot. They walked side by side, both chewing coca to ease the dizziness of walking at such high altitudes. All of a sudden, on the hillside above the riverbank, they saw a group of huge red-bark trees in full bloom. Not even Mamani, in all his years as a *cascarillero*, had ever seen such a sight. *Calisayas* usually grew in mixed groups, each one cross-pollinating with another variety close by. Yet here were fifty trees at least, all alike and grouped together in a single cluster. With their lilac-scented flowers and rich red foliage, it must have seemed to the two men that they had stumbled upon some fabulous, fiery jewel.

As the trees were in flower, it was too early in the season to pick any seeds. In any case, the mountainside where they grew was too steep and too overgrown to be easily reached. Ledger and Mamani could only look helplessly across at them, and then continue on their journey.

Some years later, Ledger thought again of the *calisayas* he had seen. By then he was in Australia, having nursed his flock of alpaca onboard ship across the Pacific with the help of Mamani's son, Santiago. The alpaca had not fared well in their new home. Moreover, having invested £7000 of his own money in the venture and lost it all, Ledger was broke. There was no choice for him but to make preparations to return to South America.

While in Sydney in 1860, Ledger chanced to read in a local newspaper of Markham's expedition and the arrival in India of the cinchona seeds. Ledger had, in fact, first heard of the expedition three years earlier, and had written to Markham, though he did not know whether he had ever received his letter. Reading once again about the search for cinchona, however, Ledger realised that Markham had never gone to Bolivia. This could only mean that he had failed to obtain the best *calisayas*. Santiago Mamani, who was shortly due to return to Peru, was despatched with a letter to his father. Ledger recalled to Manuel Incra Mamani the fifty huge *calisaya* trees they had seen in flower nine years before. Mamani should leave at once for Bolivia, Ledger instructed, and collect as many seeds as possible. In addition to the letter, Santiago took with him two hundred Spanish dollars. This was just a down payment, Ledger explained. When Mamani returned to Tacna with the seeds, Ledger or his brother-in-law would make him another generous payment. But for the moment this was all he could afford. He had not a penny left.

Another five years would pass before Ledger, having returned to Tacna, heard a knock at his door late one night in May 1865. Standing on his threshold was Mamani, who had walked more than a thousand miles from Coroico in Bolivia, sustained by nothing more than a little food and a few coca leaves. He reached under his old striped woollen poncho and drew out a small pouch. Loosening the knot that secured it, he poured a thin stream of seeds into Ledger's hand.

Mamani had left for Bolivia as soon as Santiago had

brought him Ledger's letter. There he had found the true *calisayas*, or at least trees very similar to the ones he and Ledger had seen in 1851. But the season was too advanced, the winter was fast approaching, and there were no seeds to be found. Mamani would have to return the following March, when the *calisayas* would be in flower. This he did, accompanied by his sons, who helped him cut bark from other cinchona trees while they waited for the seedpods to be ready. But in early spring there was a severe frost, and all the best seeds were destroyed. The same thing happened the following year, and again in 1863, and yet once more in 1864. It was not until April 1865 that Mamani was able to collect the seeds for Ledger.

That year Mamani and his sons watched as the old trees slowly brought forth an unusually large number of beautiful creamy white flowers, like the ones he and Ledger had seen so many years before. The branches were covered with a rough moss that was scarlet and bright silver in colour. There was no frost, and when the underside of the leaves turned a dark purple and the leaves were about to fall, Mamani and his sons collected the ripe seeds. In all, they filled two sacks that together weighed about forty pounds.

In Tacna, Ledger emptied the seeds onto a sheet. For a month he took them outside and spread them out in the sun, turning them every two hours to make sure they dried evenly. Then he divided the harvest in two, wrapped one half in chinchilla skins, placed them in a box that he covered with hide and addressed it to his brother George in London. Ledger paid Mamani five hundred dollars and gave him two mules, four donkeys and a gun as well as ammunition. If he could gather another twenty pounds of seeds, he told him, he would pay him a further six hundred dollars.

The box of seeds reached George Ledger towards the end of September 1865. He went immediately to Kew Gardens. Sir William Hooker, Richard Spruce's sponsor and the main force behind the British cinchona initiative, had just died. His son Joseph, who had replaced him, was sick at home, and George was informed that Markham was also absent. What he did not know was that only a few days before, the government had decided to put an end to Kew's cinchona activities. The time had come, Joseph Hooker had been informed in an official letter, 'when your account with this office may be closed'. The cinchona plants and seedlings that had been brought from South America by Spruce were growing nicely in the government botanical gardens in Mysore, and Kew's job, as carer to the plants that were waiting to be shipped to the colonies, was done. The unfortunate coincidence of this order and Sir William Hooker's death may explain the indifference shown to George Ledger by the very people he might have expected to seize upon his offerings with enthusiasm.

On returning from Kew, George called on Harry James Veitch, scion of the great orchid-trading family and director of the Exotic Nursery at Chelsea – just a short walk from the Chelsea Physic Garden to which, nearly two hundred years earlier, Sir John Evelyn had hurried to visit the 'collection of innumerable rarities [including] the Tree bearing the Jesuits bark, which had don such cures in quartans'. There he found a more attentive listener, who agreed to plant out a few of the seeds. On 2 November he reported to George that they were 'germinating freely'. Meanwhile, George wrote to the Under-Secretary of State for the Colonies, who, not knowing what to make of his offer of seeds of the true *Cinchona calisaya*, copied it to Joseph Hooker at Kew.

George began to panic. The seeds might deteriorate, and

he was making little progress. He visited Professor Robert Bentley, an early member of the Pharmaceutical Society and an expert on medicinal plants, giving him some of the seeds. Bentley sent them to the Botanical Society of London. George also gave some seeds as a present to the Society of Arts, who sent them on to Lady Hanbury, a relative of the Daniel Hanbury who had helped Spruce while he was in South America. Finally, Bentley put George in touch with J.E. Howard, a quinine manufacturer in Tottenham in north London. Howard's friendship with the Ledger family would last until Charles's death in 1905.

Howard was the first to appreciate the commercial potential of Ledger's seeds. Knowing that neither Clements Markham nor any of the other people who were most knowledgeable about cinchona were in London, he suggested that George contact the Dutch Consul-General. The Consul himself did not know what to make of the seeds, and passed the matter on to Professor F.A.W. Miquel, a well-known Dutch botanist who had made a study of the cinchonas introduced to Java by Hasskarl, and who was then on leave in Europe. Miquel was intrigued by George's story, and on his suggestion the Dutch government instructed the Consul to purchase one pound of his seeds. The Consul offered George a hundred Dutch guilders, with the promise of more if the seeds germinated once they reached Java.

Despite this success, George still had fourteen pounds of seeds on his hands. Howard suggested he approach a Mr J.W.B. Money, an Anglo-Indian planter and owner of extensive cinchona plantations in India who was then on holiday in London. There followed a long exchange of letters, and George eventually secured an appointment. He told Money the story of the seeds, the best in all Bolivia, taken from an area in which no European plant hunter other than Ledger had ever ventured. After considerable persuasion by George, Money finally gave in and bought the remaining seeds for £50. Yet no sooner had he done so than he began to have cold feet. *Cinchona succirubra* was the best-known variety. No one had any experience of *Cinchona calisaya*, or knew how it would fare when planted out in Madras. Fifty pounds was a large sum of money, and the thought that he might have made an ill-conceived investment was too much for Money to bear.

The very next day he made his way to none other than William McIvor, the former superintendent of the botanical gardens in Mysore who had overseen the arrival and subsequent demise of Markham's plants in India, and who was holidaying in London while waiting to take up his new appointment as the superintendent of the government cinchona plantations at Nilgiri. He agreed to take Ledger's seeds, and to give Money an equivalent amount of *succirubra* seeds in exchange. Money heaved a sigh of relief.

McIvor carried the *calisaya* seeds back to Nilgiri. After travelling many thousands of miles, and withstanding all manner of hardships, Ledger's trove was now more or less evenly divided between Britain and Holland, two nations that would fight over the commerce of cinchona just as they had fought across the centuries over so much else.

McIvor's new seeds were planted alongside the *Cinchona succirubra* that Spruce had sent from Ecuador and the ill-fated *calisaya* seedlings that Markham had despatched all too hurriedly via Southampton and the Red Sea. Perhaps Ledger's seeds had been out of the ground too long, perhaps it was the climate, but the fact remains that while those McIvor planted in India failed to flourish, those George Ledger had managed to sell to the Dutch Consul-General in London grew so well in

Java that they eventually transformed not just the Dutch plantations, but the entire cinchona industry. For when the time came to measure the quinine content of the young trees, Ledger's *calisaya* seedlings in Java would prove richer than any tree had done before. When a consignment was put up for auction in Amsterdam in 1877, the bark fetched more than five times the price of a similar consignment of *succirubra*.

While Charles Ledger's *calisaya* ensured large profits for the Java plantations and the merchants who traded on the Amsterdam exchange, and even after much debate won him a Dutch pension, for Manuel Incra Mamani, who had trekked through the forests of Bolivia and waited there through frost after frost, there was no such reward.

When Charles Ledger learned of the success of his brother George's trade with the Dutch, he wrote a letter to Mamani in Coroico, the small mountain village that overlooked the valleys where he had collected his first batch of seeds and where he now lived. Charles asked him to set out once more in search of seed in the Bolivian forest, and enclosed a down-payment against delivery of a future consignment. On his return from the forest, however, Mamani was arrested. The chief of police of the district had got wind of the fact that he was collecting seeds for a foreign buyer who wanted to smuggle them out of the country. He ordered that Mamani be imprisoned and beaten until he revealed who he was working for. But Mamani refused to betray Ledger. The police chief ordered him to be beaten again. For two weeks he was given little water and no food. Realising that he would not break, the police finally released him twenty days later. They confiscated his donkeys and his blankets, and he was forced to make his way to his village on foot. When he reached his home, his son Santiago later reported to Ledger, he could hardly stand.

His damaged organs began to bleed anew, and a few days later he died.

When Ledger's *Cinchona calisaya* seeds were fully grown, it was found that the quinine content of the bark amounted not to the normal 2 or 3 per cent, but to as much as 14 per cent. Shortly after that, botanists studying Ledger's seeds came to the conclusion that they were a previously unknown kind of cinchona. In his honour the variety was named *Cinchona ledgeriana*. Charles returned to Australia, where he would die a pauper. But his reputation is slowly being rehabilitated, and in 1994 a tombstone was finally erected in his memory in Sydney. Manuel Incra Mamani, who gave his life to help Ledger's enterprise, has no memorial of any kind.

9

The Science – India,
England and Italy

*'If you would see all of Nature gathered up at one point,
in all her loveliness, and her skill, and her deadliness, and
her sex, where would you find a more exquisite symbol
than the Mosquito?'*

HENRY HAVELOCK ELLIS, 1920

*'I really believe the problem is solved, though I don't like
to say so . . . I have hardly restrained myself . . . <u>I am
on it</u>.'*

RONALD ROSS to Sir Patrick Manson,
31 August 1897

The garrisons that the British, the French and the Dutch
had established along their trading routes evolved as the nineteenth century progressed into full-blown colonies. No longer
was the European footprint in the tropics limited to a small
defensive presence; it grew and spread as settlements expanded and the vanguard of men was swollen with the arrival
of wives and children. Yet it was not inevitable that new discoveries about tropical diseases, especially malaria, would be
made in these growing tropical colonies. Malaria had, after all,
existed in Europe for centuries. The centres of medicine were

still to be found there, and the best hospitals. But the colonies had one significant advantage over the old European nations, and that was a vast number of malaria patients. With a ready supply of infected blood at their disposal, it comes as no surprise that the most significant steps forward in our understanding of malaria and what causes it were taken by two army officers, one French and one British, one stationed in Algeria and the other in India, men whose ambition, energy and optimism made them in every sense children of empire.

Charles-Louis-Alphonse Laveran was born in Paris in 1845, but raised mostly in Algeria. He came from a long line of physicians, and both his father and his grandfather had been doctors. In 1867 he graduated in medicine from the University of Strasbourg, on the border between France and what was then Prussia, and not far from where Philippe Bunau-Varilla had been born five years earlier. The two countries, never the firmest of friends, were undergoing one of their periodic bouts of ill will, and Laveran joined the French army. He was captured at the fall of Metz in 1870, although before long he was released and posted back to his regiment. In 1878, at the age of thirty-three, the gentle doctor with the *pince-nez* was posted to Bône, the city where St Augustine had died on the Mediterranean coast of north-eastern Algeria.

The North African coast was infested by mosquitoes that had adapted to breeding in the confined waters of the oases and the irrigation waters of coastal farms. The French had introduced rice farming when they colonised Algeria in 1830, inadvertently giving a boost to the breeding of the anophelines. In the clinic where he worked, Laveran was confronted with malaria every day.

In addition to holding a degree in public health, Laveran had spent four years as an Associate Professor of Epidemic

Medicine at the French army's medical school in Paris, Val de Grâce, occupying the very chair that had been established for his father. There he had come into contact with four doctors who were working extensively with malaria. One of them, F.C. Maillot, had spent many years in North Africa, where he had developed a highly successful practice prescribing quinine. Two others had been investigating the pathology of malaria, using the newly invented microscope to examine glass slides smeared with human blood. Perhaps it was this that led Laveran to acquire a microscope of his own.

The instrument he found himself working with in Bône was nothing like the microscopes we know today – Carl Zeiss's oil-immersion lenses, which would shortly change the world of microscopy, had not yet been invented. Laveran's eyepiece was little more than a magnifying glass, but he made good use of it.

He spent the first two years after his arrival in Algeria in 1878 going over old ground, familiarising himself once more with the symptoms of malaria. He looked at the dark pigmentation that developed in certain organs of patients who had died of the disease. There was often so much black pigment in the liver, the brain or more commonly the spleen, that the entire organ turned grey. What caused this in malaria, he asked himself, that happened in no other disease?

Laveran worked on material drawn from autopsies. The usual method was to apply a thin smear of blood to a glass slide, dry it, fix it chemically and then stain it. He made a huge leap forward from the work of researchers before him when he began to examine fresh blood, which he took from his patients. Although his microscope was weak, he knew immediately that the forms revealed when he looked at slides made of fresh blood were completely different from anything that

Known originally as 'a no one from Bône',
Charles-Alphonse Laveran in 1880 became the first
person to identify the presence of the malaria parasite
in human blood.

Major Ronald 'Mosquito' Ross, whose discovery that mosquitoes were responsible for transmitting malaria would win him the Nobel Prize for Medicine in 1902, but bring him into terrible conflict with the Italian malariologist Giovanni Battista Grassi (*below*).

Wartime propaganda posters warning soldiers to keep taking their quinine and never, ever to expose themselves to mosquitoes.

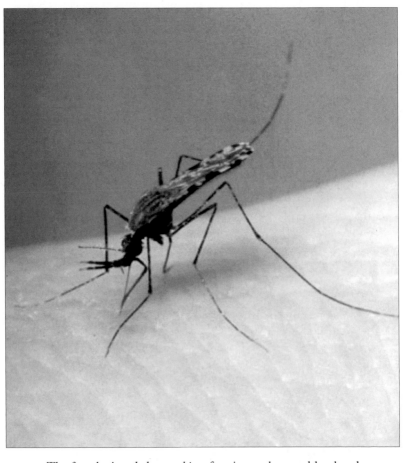

The female *Anopheles gambiae,* feasting on human blood and transmitting malaria.

had been observed before. Not only could he count off the leucocytes, the white blood cells that contained the black pigment; alongside them were clear bodies, in two basic shapes, crescents and spheres. Laveran called them the No. 1 and the No. 2 bodies. He suspected they were parasites. But how to prove it?

Laveran's lucky break came on 6 November 1880, when a feverish twenty-four-year-old soldier from the 8th Squadron of Artillery presented himself at his surgery complaining that he had been treated with quinine three weeks earlier, but once again had the ague. Examining the soldier's blood under his microscope, Laveran found large quantities of the crescent-shaped No. 1 bodies. Alongside them were a number of rounded bodies, whose surfaces were covered with filaments that were lashing and sinuously dancing on the slide: 'I was astonished to observe that at the periphery of this body was a series of fine, transparent filaments that moved very actively and beyond question were alive.'

Laveran, at that moment, became the first person to witness the parasite that causes malaria. Two days later he found more of them, as well as a number of transparent, or hyaline, ring-shaped bodies within the red blood cells. Observing them further, he noticed that the filaments, which he would later call flagella, were slightly swollen at the ends, and that now and then they would detach themselves from the spheres out of which they had emerged and swim off, like poisonous spermatozoa, stirring up the red corpuscles into a great turmoil. He also concluded that the No. 1 and No. 2 bodies he had observed were in fact the same: he watched crescents fill out and become spheres, just as he saw spheres grow hairy with filamented flagella and then despatch them off into the unknown.

Laveran described everything he had observed in a note that he sent on 23 November, less than three weeks after seeing the sick soldier, to the Académie Nationale de Médicine in Paris. His letter was greeted with disdain and disbelief. Crescents, dancing filaments, hyaline rings – how could a single micro-organism have so many guises? And why did this prove that malaria was transmitted by a parasite, when everyone knew it came from the bacteria contained in the noxious gases of the miasma? And who was this Laveran anyway? A nobody from Bône.

Laveran would spend four years trying to persuade the world that he was right. Leading Italian malariologists, including Giovanni Battista Grassi, Ettore Marchiafava and Angelo Celli, remained sceptical. They might have been more open-minded had they not felt so irritated that an important discovery had been made by a doctor from another country. The parasite – if it was a parasite – Grassi argued, had to have a nucleus, and none had been found; it also had to feed, and there was no evidence that it did. Others simply refused to believe. The Canadian Dr William Osler, probably the premier blood specialist of his day, who went on after teaching in Baltimore to become the Regius Professor of Medicine at Oxford University, for which he would later be knighted, wrote: 'Nothing excited my incredulity more than [Laveran's] description of the ciliated [flagellated] bodies. It seemed too improbable, and so contrary to all past experience that flagellate organisms should occur in the blood.' In time, though he would get little immediate credit for his work, Laveran made individual converts, who then went on to convert others. Marchiafava was the first of the Italians to accept that Laveran was right. He showed the microscopic technique to an English doctor, who in turn passed it on to

Dr Patrick Manson, who would later pass it on to Dr Ronald Ross.

Ross, who would one day become known as 'Mosquito Ross', would win the 1902 Nobel Prize for Medicine for his work on how malaria was transmitted – but not before the eruption of a bitter academic quarrel such as had not been witnessed since 1653, when Jean-Jacques Chiflet's *Exposure of the Febrifuge Powder from the American World*, in which he had denounced the new ague remedy as a Popish fraud, divided the Christian world in two.

Ronald Ross spent five years in India, where he was a member of the Indian Medical Service, looking in vain through his microscope for the *Plasmodium* parasite that Laveran had identified in 1880. On his second furlough home, in 1894, he was advised to pay a call on Patrick Manson, a kindly and avuncular Scots doctor who had made a name for himself in London as a specialist in tropical diseases, and was a great believer in the healing properties of quinine.

Manson's special interest was not malaria but filariasis, now known to be a chronic parasitic infection of the lymph system, that he had first encountered while living in China, and which he believed might be spread by mosquitoes. His curiosity had led him to become adept at using the microscope; he worked hard at the laboratory bench, and made frequent dissections of the dead. Manson had devised a number of imaginative theories about how diseases were transmitted, and at the time Ross passed through London he was still working on the idea that mosquitoes could transfer other diseases, including malaria.

As physician to the Seaman's Hospital Society, Manson came across many tropical ailments in addition to filariasis.

Malaria was one of them. As the disease had largely been eradicated in Britain by the middle of the nineteenth century, this access to a large number of malaria patients made Manson unusual among London doctors. If anyone could show Ross Laveran's *Plasmodium* parasite, it was this quiet-spoken specialist in tropical diseases.

Manson was out when Ross first called on him at his home in Queen Anne Street, London, on 9 April 1894, but he sent a note entreating him to call again, 'tomorrow if possible', and suggesting that the reason Ross had missed seeing the *Plasmodium* was perhaps because 'of the technique you employ'. He added: 'It will give me great pleasure to be of any service to you for I am quite sure you can do good work and have the patience to do it.'

Although the two men had never met, Manson had already formed a shrewd judgement of Ross's character. Photographed in his black frock-coat and high wing-collar, Ross looks every inch the confident Victorian soldier in civilian dress. He has fierce black eyes and a high, domed forehead. It is easy to imagine him pacing up and down before his drawing-room fire, deep in thought, curling his moustache in a masterful way. Manson was right: Ross could do good work, and he was certainly patient. But Ross the man was a far more complex creature than his photograph suggested.

Born in India and, like so many sons of the Raj, schooled in England, Ross had been equivocal about following his father and training to be a doctor. He had tried to become a painter, then in turn a composer, a mathematician, a poet and a novelist. Having failed to make a living at any of these, he enrolled at the age of thirty at his father's *alma mater*, St Bartholomew's Medical School on the edge of the City of London, only to dawdle once he got there, and qualify two years late.

The Ross who emerged was a competent clinician who would go on to expand his medical knowledge, taking further degrees in bacteriology and public health. But he retained a profound insecurity, a tendency to depression and a deep sense of being misunderstood, or more precisely under-appreciated, traits which combined to show themselves in a certain touchiness and arrogance whenever he felt crossed.

For years after their first meeting, Ross in his letters home to Manson would rail about the dunderheads in the Indian Civil Service who were impeding his research, the medical establishment that failed to recognise sufficiently the contribution his discoveries had made, and the foreign scientists he believed were trying to pass off those discoveries as their own. All too often, Ross regarded his own knowledge as superior to that of his colleagues. When he looked through his microscope and could not find the *Plasmodium* parasite, he did not conclude that perhaps he was using the equipment incorrectly. Rather he decided that Laveran's so-called parasites were probably nothing more than misshapen cells, even though the parasite had by then been observed for nearly a decade, and much was already known about its life cycle. Yet despite his shortcomings, which grew more pronounced as he grew older, Ross was possessed of enormous energy and drive.

Ross visited Manson many times after their first encounter in April 1894. He was drawn by Manson's deep knowledge and gentle Scottish humour; perhaps even by the older man's fatherliness. Manson, for his part, found that he liked his young acolyte more and more. And Ross, for the time being at least, seemed happy to play a subordinate role.

In his research into the spread of filariasis, Manson had made only intermittent observations of the life cycle of the parasite. He realised that it did not seem to develop much while in the blood, and thus probably needed outside help to complete its life cycle. Could the bloodsucking mosquito be that outside agent? Manson threw himself into dissecting mosquitoes. Opening up hundreds upon hundreds of their abdomens, he saw that the parasite underwent a number of subtle changes there; but he was unable to understand how, or why. He had read somewhere that mosquitoes fed only once during their short lives, an observation that was incorrect, and so failed to see the importance of the mosquito's bite. To make up for that gap in his knowledge, Manson devised his own ideas on how the parasite made the transition from human to human.

In the course of his conversations with Ross, Manson transferred ever more of his filarial theory to malaria. The parasite that caused malaria, *Plasmodium*, enclosed in a patient's red blood cells, needed an outside agent to develop, just as the filariae did. In sucking up human blood, the mosquito ingested the filariae, which then began to metamorphose. Surely the *Plasmodium* parasite would do the same? That still left a large gap in the life cycle of the parasite. To fill it, Manson posited that shortly after laying its eggs the female mosquito died, and its decayed body re-entered the chain. The female, having fed, he wrote, 'seeks out some dark and sheltered spot near stagnant water. At the end of six days she quits her shelter, and, alighting on the surface of the water, deposits her eggs thereon. She then dies, and, as a rule falls into the water alongside her eggs.' Humans would become infected by drinking the water.

Or perhaps the *Plasmodium* skipped that phase altogether. The larvae that emerged from the eggs began to eat, Manson suggested, 'and one of the first things they eat if they get the chance, is the dead body of their parent, by now soft and sodden from decomposition and long immersion'. Through this insect cannibalism, Manson believed, the parasite would enter the larvae, and thence the adult insect.

As a biological construction, Manson's theory was as ingenious as it was imaginative. The trouble was that it would, in almost every respect, turn out to be quite wrong. The analogy between filariae and *Plasmodium* would not stand up to scrutiny. Man did not acquire malaria nor the filarial worm by drinking a watery cocktail of dead mosquitoes. In the absence of conclusive evidence to the contrary, Manson clung to his theory. There was something quite Victorian about an entomological melodrama tinged with cannibalism and filial greed, but deep down, Manson wanted more than just operatic satisfaction; he wanted to be proved right. And Ross, he decided, was the man to do it.

Aboard the SS *Ballarat* on his return voyage to India, Ross honed the skills he had learned from Manson by analysing blood he drew from his fellow passengers and dissecting the cockroaches he asked the P&O seamen to catch for him.

Having two days in Bombay before he proceeded to join his regiment at Secunderabad, Ross went straight to the Civil Hospital to see if he could find any cases of tertian fever. In his notebook he described the case of a woman from Ahmedabad whose spleen showed the classic swelling of a malaria patient. Fortunately, for Ross at least, she had not yet been given quinine, which would have affected her blood sample. Alongside this entry he jotted down the task that lay ahead, as he and Manson had conceived it: to study *Plasmodium*, not in man, but in the mosquito, 'a thing which had never been yet attempted'. To achieve that, he would have to demonstrate that *Plasmodium* could survive in the mosquito, then find out what happened to them inside their insect host, how they lived and how they developed. None of this would be straightforward, but Ross was eager to get going.

On 1 May, from Secunderabad, he began his first letter to his old mentor: 'Dear Dr Manson, I write to report progress, and hope you will excuse size of paper.' For the next four months, Ross dedicated himself to his researches. He began by catching mosquitoes, but found that they ailed and died, or would not bite. Perhaps, he speculated, it was the 'awful' heat and dry wind 'like the blast of an engine furnace'. He began raising mosquitoes from larvae, feeding them on patients with parasites in their blood, then killing and dissecting them, looking for signs of parasites in their stomachs. He set up controlled experiments at different temperatures and timescales. Slowly he began to gather evidence that the parasitic crescents Laveran had seen in human blood actually turned into spheres within the mosquito's stomach, which meant they were beginning to fill out and grow. But why?

Manson, anxious that Ross's work should become better known, and worried lest he become discouraged, suggested that he write up his findings for publication. The heat in central India was growing more unforgiving as the summer wore on, and Ross couldn't use a fan in his small laboratory for fear that it would blow his mosquitoes away. Manson wrote entreating him not to give up: 'The malaria germ does not go into the mosquito for nothing, for fun or for the confusion of the pathologist. It has no notion of a practical joke. It is there for a purpose and that purpose depend upon it is its own interests – germs are selfish brutes.'

In September 1895 Ross was ordered away from Secunderabad. The city of Bangalore in southern India had been swept by an epidemic of cholera, and Ross, with his diploma in public health, was deputed to clean up the native quarter where the epidemic had begun. Organised and energetic, if occasionally overbearing, Ross found he had a gift for dealing with crises. The epidemic died down over the winter, but the following summer it flared up again. Ross once again sprang into action. He began by closing the public wells. 'We did what we could,' he would write in his *Memoirs* twenty-eight years later. 'We closed the wells in batches; we disinfected them over and over again; we told the people to boil their drinking water and provided it ourselves; we gave them hot coffee and medicine early in the morning; we disinfected backyards, drains, and rooms occupied by the dying. All in vain. The Angel of Death had descended amongst them and smote the poor wretches right and left.'

Ross worked from early in the morning until midnight. Inevitably, his work on mosquitoes and the malaria *Plasmodium* slowed. In April 1896, Manson wrote to him, 'I . . . regret very much that these [duties] have taken you away from what I look upon as your proper work – the investigation of the mosquito stage of malaria.' By October he was even more concerned – less about Ross losing interest than that the two of them might be beaten by a foreigner in their quest for a provable theory of how mosquitoes transmit malaria. 'Laveran is working up the Frenchmen,' Manson warned Ross. 'I do not hear that the Germans are moving but they will and so will the Russians. Cut in first.'

In fact, Ross was about to make a breakthrough. All his early experiments had been performed on mosquitoes of two distinct genera, *Aedes* and *Culex*. The former he called 'brindled', and the latter grey or bar-backed. Both types of mosquito can be a menace to man: *Aedes aegypti* spreads yellow fever, and was thus responsible for the single biggest problem the French encountered in Panama; *Culex fatigans* transmits the filarial worm, as well as various viruses that cause encephalitis. Neither, though, transmits human malaria. Ross had begun to suspect as much as his tour of duty in Bangalore was drawing to a close. Eager to see if another species of mosquito would answer his many questions, Ross asked for two months' leave to travel to Sigur Ghat, a valley of coffee plantations not far from the Nilgiri Hills in south-western India where, by coincidence, the cinchona seedlings Sir Clements Markham had obtained in South America had been planted out nearly thirty years earlier by William McIvor.

Instead of the right mosquito, Ross found something quite different.

Just two days after he arrived, he fell ill with pains in his liver. Within half an hour he was seized by chills. By midnight he was delirious with fever and suffering badly from aching bones. The fever lasted two hours and then broke, allowing him to sleep. The next morning he felt better, and immediately set to examining his own blood. 'I found only one small amoeboid body (quite typical and moving) in two excellent specimens,' he would write to Manson two days later. 'I don't know the species. At 9 a.m. fever began to come on again without rigor & very slightly. I slept till 3 p.m. when I woke in much perspiration & have not had fever since. Liver still hurts on a deep breath but I gave the parasites such a warm reception with quinine that I don't think they liked it.'

Ever the scientist, Ross went back over every detail of the previous few days to try to work out how he fell ill. He had slept at night under a mosquito net 'and with closed windows & doors', and had religiously drunk only boiled water and milk. He tried to calculate the volume of air he had breathed in since arriving in the malarial zone, and the number of parasites he might have ingested through drinking cool tea; but neither of these suggested themselves as a plausible cause of his illness. Manson, once he'd expressed his sympathies, was patently delighted that Ross appeared once more to be concentrating on the job in hand. 'I am very glad to hear the *Plasmodium* is to the fore again,' he wrote, 'and that you will be on his tracks so soon. Stick to him. Sooner or later he will do much more for you than sanitation work in Bangalore.'

Manson's enthusiasm notwithstanding, Ross continued to fret about how he could have caught malaria. He continued to go through the possibilities, slowly casting each one aside. The water and milk he had poured into his tea had been boiled, so there was little chance of having swallowed an infection. The air he had breathed could not have been full of parasites: Ootacamund, where he had left his family before setting off for Sigur Ghat, was not especially malarial – certainly not enough for parasites to be present in the air in large quantities.

Only later did Ross remember that in between leaving Bangalore and arriving at Sigur Ghat, he had visited the damp bottom of the *ghat* proper, 'and spent the whole day there hunting for mosquito pools and examining dew puddles', which were a well-known source of malaria. If anything led him to abandon Manson's theory about drinking in malaria from dead mosquitoes, it was the personal observation of his own sickness: 'Another five days in the jungle reversed the position of affairs in favour of the mosquito theory but only just as I was beginning to give it up,' he wrote to his mentor.

If Ross had caught malaria in an area where mosquitoes abounded, and where much of the human population carried the *Plasmodium* parasite in their blood, then surely this must mean that mosquitoes not only sucked up infected blood, but also passed it on. And when could they possibly do that, other than while biting a human? Mosquitoes were not, as Manson believed, simply a transport system that carried the plasmodium around within the environment. Nor were they, as the Italian malariologists believed, like living syringes, inoculating people as they bit them. Rather, the mosquito proboscis was a dual carriageway, a passage in both directions, in and out. To accept this meant abandoning the conventional view, long clung to by Manson and others, that the malarial germ somehow just hung about in the environment, which was not something Ross was quite ready to do yet. But he was getting there.

There was nothing to warn Ross that 20 August 1897 was to be a remarkable day in any way. By now he was back in Secunderabad, and hard at work despite the heat and the presence of colleagues whom he found irritating – 'bachelors, half-castes, men with huge families & lazy men', as he had described them in a letter to Manson a month earlier – pressing on with dissecting the brown mosquitoes in search of a sign, any sign, that might prove they carried the malaria parasite. 'Unfortunately it turned out that my mosquitoes were very few in number,' he would recount to Manson, 'but they bit well and kept well.'

Although Ross tries in his letters to maintain the calm objectivity of one scientist communicating with another, his palpable excitement begins to shine through. Victorian he might have been, but he would have been inhuman if it didn't. 'I dissected the first [mosquitoes] without any result,' he began.

On Friday 20th I had only two left. One of [these] was very carefully dissected. I noticed at once some cells in the stomach rather more distinct than the usual very delicate stomach cells. The outline was sharp but fine and not at all amoeboid, the shape spherical or ovoid, the substance rather more solid than that of the neighbouring cells. Now I am so familiar with the mosquito's stomach that these bodies struck me at once; and you may imagine how much more struck I was when, on focusing carefully I found they contained pigment . . .

It was black or dark brown (not blue, or yellow or greenish . . . Now what do you think of this? Examining next day, the cells, as I expected they would . . . showed up more clearly and could now be made out easily . . . The pigment had clustered more into clumps. In fact, apart from the pigment, the bodies looked parasitic . . .

On the 21st I killed my last brown mosquito and rushed at the stomach. The very same bodies(!) only larger, more distinct with thicker, perhaps double, outline (sometimes).

As he worked, Ross began making an inventory of his findings, knowing that if he jumped too soon to conclusions others in the field would be only too quick to shoot him down. 'Hence we have (provisionally),' he wrote, '1) The pigmented cells [that] appear to exist in malariated brown mosquitoes but not in brindled ones; 2) they are larger on the fifth day than on the fourth day; 3) they occupy only one part of the stomach; 4) they become more distinct under formaline; 5) they are sometimes intracellular, and 6) they have pigment exactly like that of the malaria parasite.'

Ever since Laveran had identified the malaria parasite seventeen years earlier, precisely by identifying the dark colour that appeared within the haemamoeba, the old name he used for the red blood cells he examined under his microscope, 'pigment' – it is the same word in French – had become a magic word, the Eureka of the study of malaria, a proof of knowledge and the key to advancement within the field.

'The remarkable thing about them,' Ross wrote to Manson of the dark cells he had found within the mosquito's stomach, 'is undoubtedly the pigment. I have never seen this before in any mosquito (I have now examined hundreds – or a thousand). It always lies in the cells and is not scattered outside them or in any other cells. In short it is as distinctive a feature of them as it is of the haemamoeba. No other bodies but these two that I have ever seen contain such pigment; and it is the same exactly in both. Wait now. You will say at once (as I used to think) that the parasite in the mosquito cannot have pigment, because this is known to derive from haemoglobin, and the mosquito has no haemoglobin. Has it not though? Why its stomach is full of it . . . Anyway there's the pigment, just like what we see in our old friend. Two points must be proved,' he concluded, looking ever ahead, 'a) that the cells are parasites; b) that they are malaria parasites.'

Ross left a few days later to pay a visit to his long-suffering wife Rosa, whom he had left in Ootacamund. He carried with him a set of mounted specimens showing his 'wonderful pigmented cells'. While there he wrote a short note to the *British Medical Journal*, then as now one of the most august scientific journals of the day. With the warning Manson had given him about foreign competition fresh in his mind, Ross was anxious to publish his discovery in case someone else 'happened to hit on the right species of insect [and] polish off the whole business'. His paper, quietly entitled 'On Some Peculiar Pigmented Cells Found in Two Mosquitoes Fed on Malarial Blood', was published in the last issue of 1897. The study of malaria had been changed forever.

Ross had been reticent about the importance of his findings in his paper, but Manson, who had forwarded it to the *BMJ* on his behalf, attached a short but enthusiastic note: 'I am inclined to think that Ross may have found the extracorporeal phase of malaria.' In private, and to Manson in particular, Ross was more excited: 'I really believe the problem is solved, though I don't like to say so . . . I have hardly restrained myself from wiring "pigment" to you, but fear you would think I had gone mad. Well I know pigment by this time. *I am on it.*'

Indeed, Ross felt so 'on it' that at the end of his historic day, even before he wrote to Manson, he turned again to poetry, writing in his small notebook:

> This day relenting God
> Hath placed within my hand
> A wondrous thing; and God
> Be praised. At His command
> Seeking His secret deeds
> With tears and toiling breath,
> I find thy cunning seeds,
> O million-murdering Death.
> I know this little thing
> A myriad men will save.
> O Death, where is thy sting,
> Thy victory, O Grave!

Ross was convinced that the final secrets of the transmission of malaria would soon be his. Perhaps it was this that tempted him to overlook the fact that although he had observed the snaky flagella that were so characteristic of the parasite in human blood, and had found the parasite developing within the mosquito's stomach three days later, he had no clearer idea – in fact, no idea at all – about what happened during the intervening period.

His unwillingness to recognise that this opened a gap in his theory also led to a certain reluctance to look around and see the true worth of the work being done by malariologists in other countries. Some, such as W.G. MacCullum, a medical student who lived outside Toronto, were, like Ross, scientists working alone with meagre resources. By contrast, Giovanni Battista Grassi, Amico Bignami and Giovanni Bastianelli worked in well-financed and well-respected departments within the medical establishment of Italy – the country that was, for many, still the heart of malaria studies, even though Italians were no longer making groundbreaking discoveries at the rate they once had done. Ross, chippy and insecure, was immensely proprietorial about his work. He found it hard to share at the best of times, and even harder to acknowledge how science often moved ahead in tiny incremental steps rather than huge leaps, with each scientist adding his or her own small stone to the main wall that had already been built by others. Ross's was a jealous nature, and this shortcoming would have serious repercussions later on.

Just as Ross's own work would later inspire others, MacCullum's research had a significant impact on Ross's thinking about how to study the transmission of malaria. MacCullum had spent the summer of 1896, the year before Ross's journey to Secunderabad, studying the *Halteridium* parasite in swamp sparrows, blackbirds and crows. He found the specimens he made from avian blood contained huge numbers of parasites that were relatively large and thus easy to observe.

In the late spring of 1897, sickness was reported among the crows of Dunnsville, Ontario. MacCullum, who was in the area, saw that the sick birds had ruffled plumage, a queer croak and an unnaturally quiet air about them. He tried to shoot one without killing it, so that he could draw blood from it while it was alive, but to no avail. Then he heard of a boy who was looking after two sick crows as pets. MacCullum leapt onto his bicycle to visit him.

Returning home after withdrawing a drop or two of blood from each bird, he slid the specimens under the glass of his microscope, and observed two parasitic forms he had never seen before: one was what he called a granular protoplasm, the other a clear form that put out waving arms, or flagella. Over a period of time he watched the flagella slowly detach themselves and begin swimming about, the same process that Laveran had observed in blood drawn from the soldier suffering from malaria in Algeria seventeen years earlier, to such mockery from the Italian Grassi, Sir William Osler and their colleagues. A number of the flagella quickly attached themselves to the granular protoplasm and, just like sperm, the strongest among them plunged its head into the sphere and finally wriggled its whole body into that organism.

What MacCullum had just observed was the fertilisation process of the *Halteridium* parasite. Could this be the answer to the question of what happened to the malaria parasite within the mosquito's stomach? Having been fertilised, the parasite multiplied, giving birth to new offspring. If this was the life cycle of *Halteridium*, could that of the malaria parasite really be different? Similarly, if it worked in this way in birds, was it not possible that the parasite operated in the same way in human beings? If so, then mosquitoes were essential not just to the transmission of malaria, but to the reproduction of the parasite.

By chance, the zoological section of the British Association for the Advancement of Science happened to be holding its annual meeting in Toronto that year. MacCullum was given the chance to read his paper on the sexual reproduction of the *Halteridium* parasite to a distinguished audience of English doctors, which included Lord Lister. They listened carefully to his dramatic discovery, and were so impressed that at the end it was Lister himself who proposed the vote of thanks. One man who was not there was Major Ronald Ross. But thanks to Mrs Manson, who typed it out for him, he was soon able to read MacCullum's paper in India, and begin adapting the experiment for himself.

Ross's own discovery of the pigmented cells within the mosquito had perhaps been announced too modestly. Whatever the reason, it had failed to have the impact he had hoped for. Before he could dwell for long on that, though, he received a telegram ordering him to proceed to Kherwara, a dry, empty spot in Rajputana, fifty-six miles from the nearest railway junction and reachable only by *tonga*, a two-wheeled horse-drawn carriage.

The secondment threw him into despair, for although it had many other palpable hardships, Kherwara was free of malaria. If there was anything that epitomised the authorities' disregard for and ignorance of the importance of Ross's work, he felt, it was this posting to a place where his precious research simply could not proceed. 'You must confess,' he wrote to Manson in November 1897, 'that it is trying to be snatched away just as one is on the point of proving a theory which one has been thinking of and working at for nearly three years; let alone being sent to a wretched place with two or three Europeans and a few itch cases and febriculae when one has had every right to expect something fairly good.'

The absence of malaria did, however, inspire Ross to take a

leaf out of MacCullum's book and try to discover how the malaria parasites developed within birds. Manson, meanwhile, had been appointed an adviser to the Colonial Office, and was about to publish his great work, *Tropical Diseases*. If Ross already felt sorry for himself, then Manson's hearty letters, while invigorating, made him deeply envious. Less than a month later, Ross was pouring his heart out again: 'Work such as mine in Bangalore would in any other branch of public service have pushed me on. In the medical service it is of no use, because the medical services have become stagnant ponds full of torpid old carp . . . The heads of service don't and won't recognise anything but seniority and red-tape rules.'

Ross complained constantly to others as well as to Manson, and was beginning to acquire a reputation as a whiner. But for once his complaining was about to pay off. On 1 February 1898 he wrote to Manson to tell him of a surprising telegram from the Surgeon-General, ordering him to take an appointment on special duty, the study of malaria.

Finally ensconced in Calcutta in a working laboratory, a luxury he had never really had before, Ross quickly advanced, despite the 'awful heat'. Using sparrows infected with the malaria parasite, he dissected mosquito after mosquito. He hoped to be able to deliver the final results of his experiments at the meeting of the British Medical Association in Edinburgh in July, perhaps even in person. Towards the end of June, he wrote to Manson: 'I think I am on the verge of another great advance.' Time after time, he observed what he called 'rods' in the body cavity of the insects he dissected. He thought they were spermatozoa, or 'naked spores', but their behaviour did not fit anything he had imagined within the life cycle of the parasite. He pushed on, and on 6 July he wrote to Manson again.

As in his earlier letters, Ross's quick-running handwriting weaves around drawings of cells and ducts and mosquito heads in full array. 'I hope this letter will reach you in Edinburgh,' he begins after brief congratulations to Manson on a new job. He moves briskly on to his own preoccupations, and his excitement at sensing that he is closer than ever to his holy grail is plain on every page even now, more than a hundred years on:

I feel almost *justified* in saying that I have completed the life cycle, or rather perhaps one life cycle, of proteosoma [a malaria parasite now called *Plasmodium relictum* that is found in birds], and therefore in all probability of the malaria parasite. I say almost, because though I think I have seized the final position, I have not yet occupied it with my full forces.

My last letter left me face to face with the astonishing fact that the germinal rods were to be found in the thorax as well as in the abdomen. Instead of the hard resisting spores we expected to arrive at – spores easily seen and followed – here were a multitude of delicate little threads, scarcely more visible than dead flagella and poured out amongst the million objects, which, under an oil-immersion lens, go to make up a mosquito. I dare say you imagined my consternation. I could not conceive what was to happen to the rods.

Well, I was in for a battle. It was, I think, the last stand – on the very breathless heights of science. I am nearly blind and dead with exhaustion!! – but triumphant. Expect one of the most wonderful things.

The rods were evidently in the insect's *blood*. By merely pricking the back of the thorax and letting the milky juice flow into a minute drop of salt solution thousands of proteosoma-coccidium-rods could easily be obtained. The question was what next?

Here Ross breaks off, only to start writing again a short while later.

I now divided my insects before dissection between the thorax and abdomen and examined each part separately. It was found that the rods were often *more numerous in the thorax than in the abdomen*; there were even cases where scarcely one could be found in the abdomen (the coccidian evidently burst some days previously as shown by their empty capsules), while numbers (4 or 5 in a field) could be detected on teasing up the thorax and head.

Here however I was brought up standing. Sometimes the rods were more common in the head, sometimes in the thorax. I went at mosquito after mosquito spending hours over each, until I was blind and half silly with fatigue. The object was to find if possible a place or structure where the rods accumulate; or to discover some further development in them. Nothing.

On the 4th however, after pulling out the head by its roots (oesophagus etc) from the thorax, some delicate structure dropped out of the cervical aperture of the latter. This proved to be a long branching gland of some sort, looking like a coil of large intestine, and consisting of a long duct with closely packed, refractive cells attached to it. I noticed at once that the rods were swarming here and were even *pouring out* from somewhere in streams. Suddenly to my amazement it was seen that many of the cells of this gland contained *the germinal rods of proteosoma-coccidia **within them***. Looking further the cells of one whole lobe of the

273

gland were simply packed with them, and on bursting the cells the rods poured out of them just as they pour out of the original coccidia.

Here Ross's nib breaks, and he splashes ink across the page. After re-equipping himself, he sets off once more.

The rods were quite unmistakeable, having the tapering, flattened and vacuolated structure peculiar to them. They are identified at once and no structures like them exist in the normal mosquito. Here they were in the cells of the gland. [The cells] have a very thin outline and contain a perfectly clear fluid, without granulations or oil-drops such as coccidian possess. The rods lay within them quite regularly and motionless except for Brownian movement. In one lobe almost every cell contained numbers of rods; in other lobes only one or two cells contained them. By the attachment of the cells to the central duct, it seemed quite easy for the rods to pass on occasion from the former into the latter.

Now what was this gland? Will you believe it, I examined two whole mosquitoes without finding it again? What with the scales, the debris of muscle etc, I could not come upon it. A third mosquito gave the same result, until I opened the *head* itself. There was the gland *attached by its duct which led straight into the structures somewhere between the eyes*. The cells were again packed with germinal rods.

I have found the gland now altogether in seven mosquitoes. In six of them the cells were packed (especially in some lobes) more or less with the germinal rods. In the seventh I could find only a small piece of gland, which was free from rods.

I still experience, however, the greatest difficulty in dissecting out the gland itself. It appears to lie in front of the thorax close to the head, but breaks so easily in the dissection that I cannot locate it properly. In the second mosquito however there was no doubt, as shown by evident attachment that the duct led straight into the headpiece, probably into the mouth.

In other words it is a thousand to one, it is a *salivary gland*.

I think that this, after further elaboration, will close at least one cycle of proteosoma, and I feel that I am almost *entitled* to lay down the law by direct observation and tracking the parasite step by step.

Malaria is conveyed from a diseased person or bird to a healthy one by the proper species of mosquito, and it is inoculated by its bite.

Remember however that there is virtue in the 'almost'. I don't announce the law yet. Even when the microscope has done its utmost, healthy birds must be infected with all due precautions.

I say one cycle. I think it is likely there is another. I continually observe that only a portion of the coccidian contain germinal rods. The rest, I now think, give rise to the blank sausage-shaped bodies . . . which I believe may be the true spores of the parasite meant either for free life or to infect grubs. Oddly enough, in old mosquitoes these bodies also get carried away into the tissues – unless they are some disease of the insect. I will attack this next.

7th. I dissected two healthy mosquitoes this morning and began by dividing up the head and the anterior third of the thorax from the middle third by means of a razor, and then carefully breaking up the anterior third. In both cases the glands were found and their ducts were traced straight into the head.

In all probability it is these glands which secrete the stinging fluid which the mosquito injects into the bite. The germinal rods, lying, as they do, in the secreting cells begin to perform their function, and are thus poured out in vast numbers under the skin of the man or bird. Arrived there, numbers of them are probably instantly swept away by the circulation of the blood, in which they immediately begin to develop into malaria parasites, thus completing the cycle.

No time to write more . . .

Manson was the first to recognise the full significance of what Ross had achieved. Despite a bad attack of gout that had left him weak and in pain, he set about making preparations to address the British Medical Association meeting in Edinburgh.

Ever unselfish, Manson wanted to be the first to tell the world about Ross's work and why it was so important. This remarkable *Plasmodium*, with all its sexual complexities, the fecundity of its cells that implanted themselves in the walls of the mosquito's stomach, the way in which each one grew into a mature individual, navigating unerringly within the mosquito's innards towards its head and into the salivary glands, where it would steady itself at the top of the insect's tubular mouth ready to be injected into the next animal host – all of this had been observed by just one man.

Manson was a warm and generous speaker, and the image he conjured up that July afternoon in Edinburgh was not so much of a new world being unveiled, as of one man who had swum through an uncharted ocean, sighted land and eventually, as a result of his efforts, gained a toehold on a continent

The life cycle of *Plasmodium*

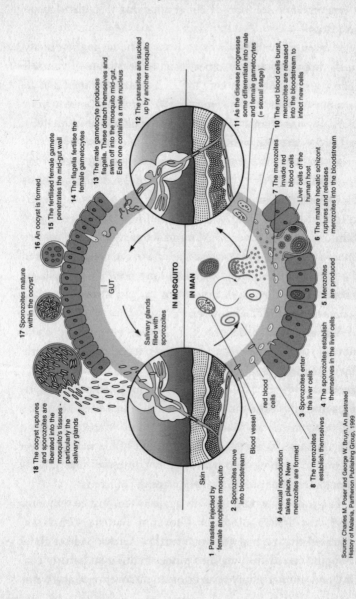

1 Parasites injected by female anopheles mosquito

2 Sporozoites move into bloodstream

3 Sporozoites enter the liver cells

4 The sporozoites establish themselves in the liver cells

5 Merozoites are produced

6 The mature hepatic schizont ruptures and releases merozoites into the bloodstream

7 The merozoites invade red blood cells

8 The merozoites establish themselves

9 Asexual reproduction takes place. New merozoites are formed

10 The red blood cells burst, merozoites are released into the bloodstream to infect new cells

11 As the disease progresses some differentiate into male and female gametocytes (= sexual stage)

12 The parasites are sucked up by another mosquito

13 The male gametocyte produces flagella. These detach themselves and swim off into the mosquito mid-gut. Each one contains a male nucleus

14 The flagella fertilise the female gametocytes

15 The fertilised female gamete penetrates the mid-gut wall

16 An oocyst is formed

17 Sporozoites mature within the oocyst

18 The oocyst ruptures and sporozoites are liberated into the mosquito's tissues - particularly the salivary glands

Liver cells of the human host

IN MAN

IN MOSQUITO

GUT

Salivary glands filled with sporozoites

red blood cells

Blood vessel

Skin

Source: Charles M. Poser and George W. Bruyn, An Illustrated History of Malaria, Parthenon Publishing Group, 1999

that had never been visited before. The meeting ended with a unanimous resolution of congratulations to both Manson and Ross.

Manson wrote afterwards to Ross: 'The fat is thoroughly in the fire and you may expect soon to hear more of yourself than your modesty may care for . . . I was determined if at all feasible to come to this meeting & properly present your work, the importance of which it is difficult to overestimate.'

Lord Lister, he went on, was there and 'is taking a very great interest in the progress of the investigation'. Present too was William Osler, then Professor of Medicine at Johns Hopkins University in Baltimore, who had been most sceptical about Laveran's early findings, but now a convert. 'I gave him one of your reports,' Manson wrote. Set on the eastern shore of Maryland, Baltimore was one of the malarial cities in the United States, and Osler had long been an enthusiastic proponent of the use of quinine at the city's main hospital, where he was the chief physician.

Manson's letter delighted Ross, but even so he could not help feeling more than a little envy. It had been their work, their moment. But only Manson had been there. As a later account of the discovery put it: '[Manson] had been scrupulous in giving Ross the credit; but still it had been Manson's announcement, in Manson's voice; and Manson received the applause. Years later Ross went out of his way to collect memories of those who had been in Edinburgh that July, as though straining still to hear the clapping of hands.'

His jealousy would flare up again when, just five months later, the Italian professor Giovanni Battista Grassi announced that he had gone one further. Grassi had set about trapping and identifying mosquitoes in different parts of central and southern Italy in an effort to discover once and for all which it was that transmitted malaria in human beings. Not only had he found that the culprit was *Anopheles*, he had also proved for the *Plasmodium* that caused human malaria what Ross had discovered about malaria in birds.

Although Grassi had been inspired directly by Ross's findings, and in writing up his own experiments showed that he had followed all of Ross's procedures step by step, he made only a brief and ambiguous reference, late in his paper, to Ross's work. This constituted the height of scientific rudeness. Perhaps Grassi, as the head of a prestigious university department and with a reputation as the premier zoologist of Italy, if not of the world, thought he could disregard the little-known Major Ross, who worked in India with neither laboratory nor assistants. Whatever the reason, Grassi's discourtesy broke all the rules of scientific good manners, and it enraged Ross. For the rest of his life he would never let pass an opportunity to insult the Italians and denigrate their achievement. Even being made the sole recipient of the second Nobel Prize for Medicine in 1902 would not be enough to compensate Ross for the slight.

Grassi's scientific reputation also suffered badly as a result of his behaviour. The truth was that after four years of research, Ross had discovered the development of one species of *Plasmodium* in one species of mosquito. The parasite on which he worked, *Plasmodium relictum*, is one of twenty-five species that infect birds alone. Others infect mice and rats, lemurs, monkeys, porcupines, squirrels, bats, lizards and snakes. There are only four malaria parasites that infect man: *Plasmodium vivax*, the most common and the most benign form; *Plasmodium ovale*, a rare species native to the west coast of Africa; *Plasmodium malariae*, which causes an extraordinarily tenacious fever; and the deadly *Plasmodium falciparum*, that still kills nearly three million people every year, most of them children.

Ross's observations of the life cycle of the parasite started with the fertilised eggs on the wall of the mosquito's stomach, and ended when the 'rods', the threadlike sporozoites, had made their way into the mosquito's salivary glands. He knew that healthy birds were infected with parasites after being bitten by mosquitoes that had previously fed on infected blood; he did not know whether the cycle was the same in human beings, nor did he know anything about how the parasite developed once in the bloodstream, either in man or in birds. Grassi would paint in one corner of that map; how the parasite developed in man, and just how quinine acted upon it, would be the work of others.

As the twentieth century dawned, the age of ignorance appeared to be drawing to a close. Yet malaria was more of a problem than it had ever been. And, despite the celebration in 1930 of the three hundredth anniversary of the discovery of quinine, a large proportion of mankind was about to be deprived of its most famous cure.

The Last Forest – Congo

'The bark is distinctly better than the bite.'

DR C.E. CASPARI, St Louis College of Pharmacy,
speaking at the 'Celebration of the Three Hundredth
Anniversary of the First Recognized Use of Cinchona',
St Louis, Missouri, 31 October–1 November 1930

Until the middle of the twentieth century, the legend about the Countess of Chinchón who was cured of a fever after swallowing an infusion of the bark of a tree that grew near Loxa was still regarded as a true story. Sir Clements Markham had been greatly taken by it, and for a long time after his visit to Peru no one had reason to doubt it. Which was why a large number of botanists, chemists, pharmacists and historians gathered together in the early autumn of 1930 at the botanical garden in St Louis, Missouri.

They were there to celebrate the three-hundredth anniversary of the first recognised use of cinchona, 1630 being the date that the Countess of the legend was said to have been cured with a dose of quinine. Many of those attending had travelled halfway across America. Others had come from Scotland, and even as far afield as Indonesia.

Among them were George and Frederic Rosengarten from Philadelphia, descendants of the Rosengarten who had supplied Dr Sappington with the quinine he used to make his fever-pills before the American Civil War. In the intervening years their company had become America's biggest quinine manufacturer, and had recently merged with another large pharmaceutical company, Merck & Co.

Present also was Dr M. Kerbosch, the director of the Dutch government's cinchona estate at Pengalengan in Java. Dr Kerbosch had taken a four-month sabbatical to journey around the world with his wife, taking in the celebrations in St Louis on the way. Dr A.R. van Linge, an elderly Dutchman who ran a company called Nederlandsche Kininefabriek, was also there. He was more accustomed to working long hours in the small Amsterdam office from which he controlled the marketing and sale of all the cinchona produced in Java, and the manufacturer of 95 per cent of the world's quinine, than to gatherings of this sort. Dr van Linge was a shy man, and felt rather nervous about the prospect of addressing three hundred delegates at a banquet. It was, he told them, the first time he had ever given a speech in public.

The trustees and board of the St Louis Botanical Garden were, for their part, anxious for the visitors to appreciate that their garden held pride of place among botanists in America, with as fine a collection of plants and a history as any more venerable botanical garden elsewhere. As part of the show they put on a display of giant chrysanthemums with huge rounded heads of rust and yellow, for which St Louis was well known, and which had been specially forced in heated greenhouses so that they would have attained their rotund perfection on the day. The visitors were shown over every inch of the garden's 1600 acres of plants and trees, in particular its world-famous collection of orchids, which had been gathered from all over the world.

The climax was a magnificent banquet at the Jefferson

Hotel. The guests, most of them unused to wearing formal evening wear, gathered a little self-consciously in twos and threes. But they soon relaxed, falling into conversation with their neighbours about the tree that had drawn them together. After dinner, George Moore, Director of the St Louis Botanical Garden and the host of the evening, pushed back his chair and stood up. 'We are met here today to glorify a plant product,' he began. And glorify it they did. Casting their minds back over the journey that cinchona had taken over the past three centuries, each of the speakers told a different chapter of the cinchona story.

They spoke of the Jesuit priests who had carried the bark to Europe in their saddlebags, of the great medical and theological disputes that set Christian against Christian when the bark began to be distributed, of the changes that its very existence had forced upon the closed and dusty world of Renaissance medicine, of the damage done to the trees that grew in the Andes, and of the ambitious schemes put forward by Sir Clements Markham and the directors of Kew Gardens to grow cinchona in India, so that Britain's burgeoning colonies could have a plentiful and cheap supply of the only drug that was known to cure the fever. As the speeches moved on towards the end of the nineteenth century, the focus shifted away from botany and into the realms of chemistry, pharmacy and, inevitably, industry, as the harvest of quinine moved from field to factory. Whatever their particular speciality, the quinine delegates who gathered in St Louis on that perfect autumn evening had every reason to be pleased with what they and their predecessors had achieved.

It was the British Army's Macedonian campaign during the First World War that had finally proved, even to the most sceptical commander, how important it was that soldiers who were expected to fight in areas close to where mosquitoes bred in summer should be treated with a malaria prophylaxis if they were to remain healthy.

The British Army's medical expertise in 1914 had come a long way since the Walcheren expedition of a century earlier. Yet, despite the fact that Major 'Mosquito' Ross, by now a colonel, took part in the Greek campaign, there was still much the authorities didn't know about mosquitoes. Of the four species of *Anopheles* mosquito that are known to spread malaria in Macedonia, only two were considered at all dangerous: *Anopheles maculipennis*, which thrives at sea level, and was to be found in the Struma and Vardar valleys and west of Salonika, in the deltas of the Vardar and Galiko rivers; and *Anopheles superpictus*, which breeds at higher altitudes. At the start of the war, Britain and France were reluctant to send troops to the Balkans, preferring to let the Serbs fight the forces of Austria-Hungary alone. But when Bulgaria threatened to attack Serbia in September 1915, the Allies had no choice but to enter the conflict. They decided to establish a new base at Salonika, from which they hoped to block the German advance into northern Greece.

Unlike the British generals at Walcheren and their American counterparts at Vicksburg, the commanders of the Franco-British expeditionary force were well aware of the threat of malaria. As a result they delayed their arrival until October 1915, when it would be too cold for the mosquitoes to breed. Army camps and field hospitals were set up in the hills above Lake Langaza, and the soldiers began to advance up the Struma and Vardar valleys. The cold did not last, though it was not until early the following summer that the Allies realised their delay had been futile, and that their men were prey to both types of malaria-bearing mosquitoes. The numbers admitted to hospital quickly reached proportions not seen since Walcheren: 4500 in July 1916, 7500 in August, nine thousand by September. The following year proved even worse, with thirty thousand cases in September and October alone.

Colonel Ross was drafted in to formulate a plan of action. He proposed brigades that would drain the marshes, clean up streams and eradicate mosquito larvae. Nothing seemed to work. By the end of 1917, a further seventy thousand soldiers were suffering from malaria. The field hospitals were overwhelmed, and as it had done at Walcheren, the army began sending the worst cases back to Britain for treatment. By October 1918, when the end of the war was just a month away, there had been 162,000 admissions out of a force of 160,000 men. Some men had fallen sick with malaria twice, or even three times, each time being readmitted to hospital. Yet, despite these huge numbers, only 821 men died from malaria during the whole campaign. Troops who caught malaria were promptly treated with quinine, and were not permitted to stop taking it until they appeared to have been cured. As part of German reparations after the war, the Treaty of Versailles forced Germany in Annex VI to hand over to the Allies a quarter of its production of salts of quinine.

The First World War had made quinine many fresh con-verts. By 1918 demand for the drug had vastly increased, and the world price of quinine, though far lower than it had been in the 1860s, had more than doubled. Such prices, though, were unsustainable. In peacetime, the military would no longer be a major purchaser, and there was bound to be a reduction in price once the war was over.

In 1918 the quinine producers in Java signed an agreement

with the three remaining factories in Holland, as well as the plant in Bandoeng in Indonesia, that would help guarantee a decent price for planters while also ensuring adequate supplies. To the Europeans, the agreement, which was known as the Second Quinine Convention (a similar arrangement had been in place before the war), had been undertaken in the interests of stabilising the market for all concerned. However, it met with considerable opposition in America, where it was regarded as a restrictive monopoly. Despite that, most of the foreign visitors to the St Louis meeting in 1930 were confident that in the long term the Americans would come around to the European point of view that a stable, controlled market was in the interests of producers and consumers alike.

What they could not know, of course, was that in less than a decade a new war would break out that would involve intense fighting all over the tropics, in the course of which hundreds of thousands of soldiers would fall sick with malaria. Although many of them were treated quite successfully with the drug, the Second World War would come close to destroying the quinine industry altogether.

When Great Britain and France declared war on Germany just two days after Hitler's invasion of Poland in September 1939, the potential threat of malaria was far from the minds of the Allied leaders. Yet in the six years that followed, European, African, Asian and American soldiers would fight over more tropical and malarious terrain than in any previous conflict. From Senegal to Singapore and beyond, soldiers on all sides would fall prey to malaria in numbers not seen since the British had established garrisons that virtually guaranteed death within a year on the West African coast more than a century earlier. Along a broad equatorial belt, tropical seasonal rain, warm temperatures, year-round humidity, pools of water stagnating in tank tracks and coconut shells that allowed mosquitoes to breed with impunity, together with a heavily infected local population on which they could feed, would create perfect conditions for an epidemic of the deadly *Plasmodium falciparum* or the recurring *Plasmodium vivax*.

General William Slim, whose British, Indian and African soldiers of the 14th Army were responsible for repelling the Japanese invasion of Arakan in 1944 and for the reconquest of Burma the following year, reckoned malaria was his biggest problem after supplies – more of a danger even than combat. In 1943, over 60 per cent of his men succumbed to the disease; among the forward troops the figure was far higher. More than half of those who caught malaria would fall ill with it several times within the year.

Some of the senior French and British medical authorities had witnessed at first hand the Macedonian campaign of a quarter of a century earlier; the younger ones had read about it. The most far-sighted realised that an efficient method of treating the sick was essential. Yet that did not stop many of the Allied generals from growing impatient with attempts to ensure well-ordered medical assistance in the field. As one senior American commander at Guadalcanal in 1942 barked: 'We are here to kill Japs, and to hell with mosquitoes.'

Despite that, military doctors knew that whatever success they might have in killing mosquitoes – and eradicating them in numbers that would make a real difference was unlikely – one thing was certain. Troops – whether French soldiers in West Africa, British in Singapore and then in Sicily, Germans in the eastern Mediterranean, Japanese in China or Australians and Americans in the south-west Pacific – would die in their thousands without adequate supplies of quinine.

The quinine situation had changed radically, however,

since the balmy autumn evening when quinine growers and manufacturers had gathered in Missouri in 1930, confident that their industry had achieved a maturity which meant it could not be threatened. When the German army invaded Holland on 10 May 1940, reaching Amsterdam just days later, it quickly set about securing essential machinery and supplies. The quinine from Java that was stored in warehouses in the city was requisitioned and sent to Berlin. There was nothing for the directors of the Kinabureau, the Amsterdam-based agency that managed the Dutch quinine industry, to do but to hand control of the plantations in Java over to their managers in Bandoeng, and hope for the best. Less than two years later, though, Java itself fell to the invading Japanese army. At a stroke, virtually the entire world supply of quinine was now in enemy hands.

The Japanese were careful to keep their German allies well supplied. In 1995 the wreck of Japanese Imperial Navy submarine I-52 was found almost intact seventeen thousand feet down on the seabed in the mid-Atlantic. It had been designed not as an attack submarine, but as a cargo vessel capable of carrying hundreds of tons of war materiel. The submarine would travel underwater by day and cruise on the surface at night while recharging its batteries, thus avoiding Allied detection.

The sub had left its base at Kure, not far from Hiroshima, in March 1944, carrying two tons of gold, 228 tons of molybdenum, tungsten and tin, and fourteen experts from the industrial company Mitsubishi. They were to meet some German counterparts in occupied France in order to exchange desperately needed German technology for a supply of raw materials that the Nazis were equally anxious to obtain. On its way to the French coast, the submarine made only one stop, in

Singapore, where it refuelled and loaded fifty-four tons of raw rubber and three tons of quinine before heading towards the Indian Ocean. I-52 rendezvoused with a German submarine in the mid-Atlantic on 23 June, and picked up a German pilot who would guide it into port at Lorient in the Bay of Biscay. Within hours of the rendezvous, though, the American escort carrier USS *Bogue* had picked up a signal from the submarine and despatched a torpedo bomber to attack and sink it.

Malaria, though a nuisance, was something European troops were familiar with. For the Americans, who had rarely fought abroad, the experience would be quite different. At the start of the Second World War, only twenty-four doctors who were both fit for military service and possessed some sort of training in tropical medicine could be found in the whole of the United States. As a result, thousands of young American soldiers would die when they were sent to fight the Japanese in the South Pacific in 1942.

They had their first taste of the dangers that lay ahead in Bataan. Three days after their attack on Pearl Harbor in December 1941, the Japanese were again on the offensive against American troops, this time in the Philippines, which America had controlled since 1898. Brave, numerous and ruthless, the Japanese soon overpowered the American garrison, and General Douglas MacArthur, the commander of the South Pacific forces, ordered a swift withdrawal across the bay of Manila to the 450-square-mile Bataan peninsula. Mountainous and covered in thick jungle, Bataan was well suited to a defensive battle; but medically it proved a nightmare, with too many people to care for and too few skilled doctors and nurses to look after them, a landscape that made evacuations difficult, and, worst of all, an epidemic of mosquito-borne *falciparum* malaria which would affect virtually all the 100,000 soldiers and civilians who found themselves squeezed onto the peninsula.

Quinine was in short supply. In January 1942 the prophylactic dose of ten grains a day had to be halved, and a few months later any pretence of trying to prevent the disease was abandoned, and the remaining drug saved to treat those who were sick in hospital.

The Japanese, who were waging war in the tropics for the first time, found themselves in an equally difficult situation. Their leaders had planned for the Philippines campaign to be fought according to a rigid timetable, and although the Japanese troops on Bataan should theoretically have had access to all the quinine they needed, in fact there was only enough of the drug to last them a month. By February 1942, three months after the campaign began, just three thousand of the original fighting force of fifteen thousand Japanese soldiers were still fit, the remainder having succumbed to malaria, dysentery and beriberi. A Japanese interpreter later told American prisoners that in some units of the 14th Army the unfit rate had reached 90 per cent, with many deaths, and that Sternberg General Hospital in Manila was packed with sick Japanese soldiers from Bataan.

Whatever casualties the Japanese suffered, they could replace their sick men with fresh, well-trained troops who had fought in Malaya. For the Americans there was no such option, and many of the soldiers who had been evacuated from Bataan before it fell in April 1942 found themselves launched four months later into a ferocious attack on the island of Guadalcanal, in the Solomon archipelago, where the Japanese were building an airfield from which they planned to block the sea lanes between America and Australia.

The loss of the Philippines marked the last mass surrender

by American forces to the Japanese, but it took a long time, and heavy losses, before the tide of war turned decisively in the Americans' favour. Guadalcanal was the start.

'Guadalcanal,' an official history remarks, 'was not a picturesque South Pacific island paradise.' Hot and humid, its coastal lowlands and mountain valleys were loathed by all who fought there. Surprised by the unexpected American attack in August 1942, many of the Japanese soldiers fled the airstrip they were building, leaving behind meals that had been cooked but not eaten, and hot mugs of tea still waiting to be drunk. The Americans' supply system was plagued by confusion and inadequacy: some Marines arrived on the island unfed, as their transport had run out of all food but soup. Had it not been for the large quantities of enemy rice that fell into their hands, many American soldiers would have gone hungry.

Even those who started out well found it hard to stay fit. The heat was unbearable. A short march left men 'staggering, gasping from the sheer airlessness and almost senseless from exhaustion'. There were too few salt tablets available to the men, who were sweating profusely, and any fresh food they had quickly spoiled in the heat. Dysentery, fungal infections and skin diseases were rife. Everyone was dirty and wet, and raw, open sores soon broke out between the toes, in the groin and between the buttocks. That was before the malaria started.

Just as had happened at Walcheren, the disease broke out in the third week of August. Its progress in the 1st Marine Division was rapid: nine hundred cases before the end of the month, 1724 the next, 2630 in October. By mid-December the division had counted more than 8500 hospital admissions for malaria alone. Ninety per cent of its men caught the disease, most of them *falciparum* cases, and many fell sick with it more than once.

The experiences of Bataan and Guadalcanal persuaded the American government that malaria had to be fought in a number of different ways. In the long term, synthetic chemical drugs that were already being developed would be used with increasing regularity. Mosquito-eradication campaigns, especially after the insecticide DDT became more available, would also be more common. In the meantime, the country needed all the quinine it could lay its hands on.

In January 1943, five months after the invasion of Guadalcanal, Donald Nelson, President of the American War Production Board, recommended that a national pool of quinine stocks be established to ensure that the military had all the febrifuge it needed. The pool would be the direct responsibility of the American Pharmaceutical Association, which was told that 'each donation will be a direct contribution to winning the war'. It was exactly the sort of patriotic, all-shoulders-to-the-wheel effort at which Americans excel, and the project's goal was set at collecting 100,000 ounces of quinine.

Pharmacists in non-malarial areas of the country were asked to send all the powder, tablets and capsules of quinine, quinidine, cinchonine, cinchonidine and quinine salts they had. It didn't matter if stocks were old, so long as they were dry. Within days, packages were beginning to arrive at the American Pharmaceutical Association. The lower floor of its headquarters was devoted to the pool, and volunteers went from shop to shop ensuring that pharmacists responded immediately. Robert Rodman, one of the editors of the Association's journal, went on the radio to explain the pharmacists' role in helping to save American troops who were suffering from malaria. Supplies poured in from all over the country, President Roosevelt himself forwarding a gift of one hundred pounds of quinine that he had been given by the President of Peru during a state visit.

In five months, over three tons of quinine were collected, the equivalent of 9.6 million five-grain doses. By October, after ten months, almost five tons had been amassed, more than thirteen million doses in all. The figure comfortably exceeded the Association's target, but donations were slowing down, in large part because America was simply running out of quinine. Something more would need to be done. With such a major proportion of the world's quinine supply off-limits to the Allies, it was imperative, as it had been in Charles Ledger's day, to locate a source of seeds from which to start new plantings. The US authorities thought they had just the man for the job.

Arthur Fischer had first gone to the Philippines in 1911 with a degree in forestry and a reserve army commission. In 1917, as Director of the Bureau of Forestry in the islands, he had ambitions to develop a local source of quinine to treat the high incidence of malaria among agricultural workers. But the Dutch, like the Peruvians and Bolivians a century earlier, kept a tight embargo on all shipments of seeds and plants from their Java plantations.

Fischer, though, was persistent. He enlisted the help of the Governor General of the Philippines, General Leonard Wood, who had trained as a doctor at Harvard, and arranged for an English sea-captain to smuggle seeds from Java. The seeds arrived in two Horlick's malted milk jars, and the captain was paid $4000, the money coming from General Wood's discretionary fund. By 1927, Fischer had a plantation of the quinine-rich *Cinchona ledgeriana* up and growing on the Philippine island of Mindanao, and nine years later the first shipment of bark arrived at a small new factory in Manila. By the time war broke out, the factory was producing 2.5 kilos of quinine sulphate a day.

Along with thirty-eight thousand acres of Dutch plantations in Java, the factory in Manila was seized by the Japanese at the start of the Philippines campaign early in 1942. Within a short time, malaria cases on Luzon, east of Bataan, were increasing at an alarming rate. US Army correspondence reveals just how serious the situation had become: 'There are now three thousand cases of malaria in the Luzon Force,' the medical officer in charge of the area reported to his superior officer, the Assistant Chief of Staff, on 23 March, two months after the start of the campaign. '20 per cent of fighting units are not effective due to malaria and 50 per cent of some units have subclinical malaria and are considered potentially ineffective . . . The daily admission rate for malaria now lies between 500 and 700 cases and in the absence of quinine this rate can be expected to increase.

'There is sufficient quinine in the Philippine Medical Depot to very inadequately treat 10,000 cases . . . this will be exhausted in three to four weeks. When stocks are exhausted, a mortality rate of 10 per cent in untreated cases can be expected. Those who do not succumb to the disease can be classed as non-effective from a military standpoint due to their weakened and anaemic condition. Blood-building foods and drugs are not available.' Radio messages were sent to Melbourne, the nearest supply base, for fresh stocks of quinine, but without success.

Arthur Fischer, who was interviewed shortly before his death in 1962 by Margaret Krieg while she was researching her book *Green Medicine: The Search for Plants that Heal*, recalled that on 18 March 1942, by which time he was serving as a colonel in the US Army, he wrote to the Assistant Chief of Staff to outline a plan that he had been considering for some weeks. 'At the quinine plantation at Katoan, about forty kilometres west by south of Malaybalay, Bukidnon, under the supervision of Forester Altamirano,' he wrote, 'there should be several thousand trees over three years old from which bark can be harvested and sent to Bataan by plane. This bark should run from 7 to 9 per cent alkaloid content and can be used by grinding the bark and making an infusion with boiling water as a tea. The seed of *Cinchona ledgeriana* should be harvested and sent by plane to the United States for shipment to Brazil to start American sources.'

General Jonathan Wainwright, the American Chief of Staff in the Philippines, called Fischer immediately for a first-hand report on how this could be achieved. He quickly gave his approval. Fischer was suffering from blood poisoning as a result of a wound in his arm. He was feverish and in considerable pain from recurring malaria. His weight had dropped from its usual 150 to just ninety-six pounds, but he did not let that deter him, and left Bataan immediately for Mindanao to supervise the harvesting. With local help, he stripped bark, ground it up in a corn grinder, and packed it into old petrol drums. A first cargo was despatched, but the plane carrying it was shot down. Meanwhile, a Catholic priest with a smattering of chemistry improvised a laboratory equipped with two old bathtubs and some ether and sulphuric acid begged from a nearby hospital. Within days, though, the weakened American force on Bataan was obliged to surrender, leaving Fischer, the priest and a small guerrilla force to fight a rearguard action against the Japanese on Mindanao on their own.

The group's one thought was how to escape with a supply of *Cinchona ledgeriana* seeds. General MacArthur, who had ordered the withdrawal from Bataan and was by now stationed in Australia, ordered three B-17 Flying Fortresses to be despatched to the island to bring back the American military personnel. The first two were shot down, but the third made its escape from Mindanao. Colonel Fischer was aboard, and so was a large tin of *Cinchona ledgeriana* seeds. It was the last Allied plane to leave Mindanao before the peninsula was overrun by the Japanese.

From Australia, Fischer's seeds were flown back to the Americas, from whence the species had come seventy years earlier. After being carefully germinated in the US Department of Agriculture's laboratories at Glen Dale, Maryland, four million cinchona seedlings were sent to Latin America. One of the larger plantations was started in Costa Rica, where ten thousand acres of land had been leased to grow cinchona. Other seedlings were sent to Ecuador, one of the very countries from which they had originated.

Although the war was over by the time Colonel Fischer's trees were ready to harvest, one of his youthful dreams – to make quinine available in underdeveloped countries as a 'good neighbour policy' – had been achieved. Echoes of his philanthropic ambition would be seen fifty years later in a cinchona project in Africa. For his wartime services, Fischer was awarded the United States Order of Merit and the Distinguished Star of the Philippines.

The fact that it had been so easy for Germany and Japan to annex virtually all the quinine that the world produced, and so difficult for the Allies to get any of it back, proved to the American military in particular that the time had come to pursue other ways of curing malaria. No longer was a tree that grew far away going to be the answer.

What was needed was a drug that was easy for manufacturers to produce and for doctors to administer. The *Plasmodium vivax* strain of malaria was particularly prone to recur, but taking quinine over long periods caused many unpleasant side-effects. Patients complained of ringing in their ears, vomiting, deafness and acute headaches. In their search for an anti-malarial drug that could be manufactured artificially scientists alighted on a compound that had first been synthesised in Germany in 1930. Called 9-amino-acridine, it was licensed under the name Atabrine.

Atabrine was one of several anti-malarial drugs developed by scientists working in the dye industry. For more than a century, ever since the French chemists Pelletier and Caventou first synthesised quinine, scientists had been trying to find a way to manufacture synthesised quinine commercially. In 1834, Friedlieb Runge, a German chemist, succeeded in manufacturing quinoline, an organic-based compound, out of coal tar, though the manufacture of quinine defeated him. Two decades later, an energetic eighteen-year-old British chemist, William Perkin, put his mind to the same problem. Perkin too was foxed by quinine, though he did manage to formulate mauveine, a distinct colouring agent that became the first of the aniline dyes and launched the craze for pale purple.

In the 1920s, after their experience of trying to defend their East African colony Tanganyika during the First World War, the Germans turned once more to trying to find a synthetic substitute for quinine. Scientists working for IG Farben, part of the Bayer dye works in Elberfeld, began testing thousands of compounds. In 1926 they discovered one that appeared to kill off malaria parasites in human blood. Named Plasmochin, it was also discovered to be highly toxic, and was discarded. But the search continued.

Six years later, the scientists at Farben alighted on the compound which they named Atabrine. Others followed, but it was Atabrine that endured, because, although it turned patients' skin yellow, it appeared to be almost identical in its action to quinine, only with even better results. It remained in the blood for at least a week after the first dose, and could thus be used as a prophylaxis.

Farben sent examples of its new compounds to its American sister company, Winthrop Stearns, and in 1938 the US Army began testing Atabrine on soldiers in Panama. In the autumn of 1942, soon after the landing on Guadalcanal, Atabrine was issued to troops in the south-west Pacific 'by the roster', as the Navy Medical Department described it later, with one tablet every day except Sundays.

The unpleasant side-effects made the drug highly unpopular among the soldiers. Some suffered from diarrhoea; others could not take it without vomiting; nearly all found that it turned their skin a violent yellow. But it was only when rumours began to circulate that Atabrine caused impotence that the troops began refusing outright to take it. They threw their tablets into latrines, or hid them under mattresses; anything rather than swallow it down.

With nothing better to prevent their men from catching malaria, the authorities did their best to stress the positive aspects of taking Atabrine regularly. The truth was that there was no perfect anti-malarial available. Nor would there be until after the fall of Tunis in May 1943, when a group of French doctors suggested the Americans prescribe a drug called Sontochin, which they found had worked well against malaria in North Africa.

Sontochin, the Winthrop researchers recalled, was one of the many compounds that had been passed on by their

German sister-company IG Farben before the war, and had been gathering dust on Winthrop's shelves through the intervening decade. Another, which soon became much more popular, was a similar compound named Resochin, better known nowadays as chloroquine.

Researchers analysing chloroquine found that the drug was a solution to many aspects of the malaria problem. It was fast-acting and easy to administer, and its side-effects were relatively mild. It takes less than thirty minutes for the malarial sporozoites that are injected into the bloodstream when a person is bitten by an infected mosquito to make their way to the liver. The parasite needs to absorb some of the proteins that exist within the human liver before it looses itself back into the bloodstream in search of red blood cells. It ensures an adequate supply of protein by wandering from cell to cell within the liver. Once it has found what it wants it simply replicates over and over, many thousands of times, until as an army it re-enters the bloodstream in a form that is able to invade the red blood cells that carry oxygen to various parts of the body.

Once it succeeds in piercing the outer wall of a red blood cell, the parasite once again sets about finding and synthesising the proteins it needs. It uses them to form tiny protrusions, called 'knobs', on the surface of the red blood cell. The effect is to anchor the cell to the wall of a capillary, like a boxer against the ropes, and prevent it from being carried to the spleen, where damaged red blood cells are normally destroyed. Once inside the red blood cell, the parasite can keep changing the configuration of new surface proteins so that the body's immune system doesn't have time to identify each new variant. By forcing the red blood cell to change its disguise over and over, the parasite effectively hides its kidnapped host from the immune system for just long enough to absorb the iron that the haemoglobin contains, and which the parasite needs in order to reproduce itself.

If a patient with malaria were to remain untreated, the parasite would reproduce over and over for as long as forty-eight hours, when the wall of the haemoglobin, or red blood cell, would be ruptured, and the newly multiplied parasites released into the bloodstream to search for more red blood cells to invade. In some cases, the parasite then accumulates in the capillaries of the brain, causing cerebral malaria and, within hours, death. Meanwhile, the wreckage of the spent red blood cell on which the parasite had been feeding would be cleared by phagocytic cells, whose job it is to carry away spent cells to the spleen, causing it to swell and turn black. It is this that gives malaria patients the hard spleens – what early physicians knew as 'ague cake' – that are so characteristic of the disease.

The development of chloroquine was able to stop the parasite developing, and to explain for the first time how quinine, to which it is closely related, actually worked.

If a patient had been taking chloroquine regularly as a prophylactic, the parasite that had reproduced within the liver would find, once it reached the red blood cells, that the drug had got there first. The haeme – or iron-producing part of haemoglobin – is actually toxic to the malaria parasite. But it has learned to render it harmless by converting the haeme into something called haemozoin, the black pigment that colours the spleen. What quinine and chloroquine do is to block the parasite from being able to convert the haeme into haemozoin, so that instead of being made harmless it remains poisonous to the parasite. Trapped within its toxic prison and unable to develop any further, the parasite soon dies.

Easy to administer and relatively cheap to produce, chloroquine had the advantage of causing far fewer side-effects than quinine. Thus it appeared to be what American researchers like to call a 'magic bullet'. Chloroquine, at first, was effective against all four species of the parasite, including the deadly *falciparum*. The quinine tree, it seemed, could be consigned to history.

The discovery of chloroquine came too late to help the Allied soldiers who took part in the war in the Pacific or the landings in Sicily. But news of its effectiveness spread rapidly. Among the British and French colonies in Africa and Indochina it was soon the drug of choice.

In Italy, meanwhile, a massive campaign was under way to repair the damage that the tanks of the German Panzer divisions had wreaked upon the elaborate constructions put up through the ages to drain the marshes of the southern Italian countryside. At the same time, the Italians were keen to begin trying out a Swiss-made discovery which, it was hoped, would eradicate the mosquito from the Campagna Romana forever. The new pesticide was called DDT, a killing compound that proved so successful in eradicating disease-carrying insects that it allowed man for a while to believe that he had become invincible. The eventual eradication of malaria, went the mantra, was inevitable. It was only a question of when.

In time, though, DDT would arouse enormous controversy about its long-term effects on other animals. Chloroquine for some years remained free of taint, but as early as 1961 doctors treating malaria patients in Cambodia, Thailand and Vietnam began to notice that it was not working as well as it had done. It was as if the malaria parasite refused to be outdone, and had redoubled its efforts to find a way around the onslaught of the drug; which is exactly what had happened.

The parasite had developed a mutation that rendered it resistant to chloroquine.

Resistance could be seen when, despite repeated treatment at different dosages, the blood still showed the presence of live parasites. Why wasn't chloroquine working any more? Many theories have been put forward for this change, but the most plausible is that the resistant parasite has somehow managed to block the uptake of chloroquine into its system. Why has this happened? No one is quite certain. What is clear, however, is that over the past forty years resistance to chloroquine has become so widespread that in many countries it is now likely to be of no help to a doctor treating a malaria patient. In at least ten Eastern and Central African countries, including Kenya, Rwanda and Congo, prescribing it has become illegal.

Chloroquine's long-term failure has had two salutary effects. First, it has shifted the perspective of doctors and researchers by proving to them that some drugs may be effective for only ten years or so, and it has forced them to consider using combination treatments to attack the parasite on several fronts at once. Thus, a patient who falls ill with malaria and is fortunate enough to be able to afford the best treatment will find that, depending on where he is and which species of malaria he has, he is likely to be treated with aminoquinolines, such as chloroquine, in conjunction with other anti-malarial drugs – such as anti-folates, quinoline-methanols and phanthrene-methanols, all of which act on the parasite at slightly different points in its reproductive cycle – and even with an antibiotic.

Within this vast new cocktail of cures is one drug that is being regarded with new deference. Cheap to produce and easily manufactured in the developing world, it has shown itself to be highly effective as a treatment for the most virulent attacks of *falciparum* malaria. If a patient with this strain of malaria is treated in the world's most sophisticated hospitals, this cheap medicine will be the drug of choice. Even more remarkable is the fact that the malaria parasite has not yet developed more than a token resistance to it. Known for centuries as a treatment against fever, it is none other than natural-growing cinchona, the Jesuits' cure.

To find the last remaining plantation of Charles Ledger's *Cinchona ledgeriana*, you need to travel not to Java, but to the heart of Africa, to the city of Bukavu on the Rwanda–Congo border.

Bukavu has never been a neat town, though for a while it tried to tame its natural tropical exuberance. By the edge of Lake Kivu, wealthy Belgian *colons* used to build white stuccoed holiday villas which they decorated with a particularly Mediterranean plumage of plumbago, scented jasmine and tall green cypruses, in an effort to pretend they were elsewhere.

Five fingers of land spread down from the town into the lake. The green hillsides are wrapped in a misty tulle at dawn, and you can see why the view reminded some of the earliest European visitors of the south of France. There can surely be no other reason why this bloodstained, dusty metropolis used to be called the Monte Carlo of Africa. My grandparents reached Bukavu during their trans-African trek in 1929; my father too has been there. During his last visit, in 1990, he sat up late watching Germany beat Argentina in the final of the World Cup. The Central African night was warm, and in his excitement my father never noticed the mosquitoes streaming in through the open windows of his hotel room towards the bright television screen. Within days he had fallen sick with malaria. A quiet-spoken African doctor treated him with locally grown quinine sulphate tablets produced by a company called Pharmakina. It was to visit this company that my father and I travelled to Bukavu in the summer of 2002.

Nearly a decade earlier, the Hutu tribesmen who live just across the river from Bukavu had embarked on a murderous rampage, killing hundreds of thousands of their fellow Rwandans, the Tutsis, in a genocidal mania that has never really been explained. Many of the Hutu *interahamwe* militias, who spearheaded the genocide, fled, still wearing their blood-stained clothes, into Bukavu and beyond into the forested hinterland of what was then Zaire. There they joined forces with soldiers loyal to the Zairean President Mobutu Sese Seko against the Tutsi-led government that took power in Rwanda. After Mobutu was overthrown in 1997 they fought for his successors, Laurent Kabila and his son Joseph. After some years, Rwanda and the Democratic Republic of Congo, as Zaire has been renamed, signed a peace agreement.

On the ground we found that little had changed. Bands of rebels controlled much of the land around Bukavu. Their lorries, collecting sacks of niodium, germanium and colomo tantalite – the spoils of raids on mines that provide essential materials for manufacturing the capacitors that operate mobile phones – are often the only vehicles on the roads in eastern Congo. The rebels terrorise the local population to keep it quiescent, and make no effort to rebuild or replant anything. The acres of tea and coffee that once formed sculpted hills for miles beyond Bukavu are completely abandoned, with the result that the forest is drawing ever closer to the town. AIDS, tuberculosis and malaria are rife. Babies born HIV positive are particularly vulnerable to these diseases, and the malaria parasite most commonly found here is *Plasmodium falciparum*, the deadliest form of the illness. Two out of three children catch malaria, and many die. Most of the local people refuse to venture far from their villages; they grow what they need for their families around the huts they live in, and even that is often stolen by the rebels.

At the top of the town, overlooking the lake, a long white wall encloses the Pharmakina factory. Above the gates you can see the swollen head of a *Ficus thonningii*, the giant African fig tree that is the centre of so many of the continent's legends. Like a cat with nine lives, Pharmakina is something of a survivor. It is now the biggest employer in the region, the only company to maintain roads, run schools and finance clinics and hospitals. Pharmakina is also the owner of the world's largest surviving cinchona forest.

Congo's cinchona plantations began in 1933 when the heir to the Belgian throne, Prince Leopold, paid the country a visit with his wife. The couple brought with them a tin of cinchona seeds, a gift that had been presented to the Prince's father, King Albert, by the Dutch Queen Juliana. Offspring of the same seeds that Charles Ledger and Manuel Incra Mamani had collected and smuggled from Bolivia, the tin of *Cinchona ledgeriana* had been sent to Queen Juliana by Dutch planters from Java.

The seeds were distributed to plantations around the region of Bukavu. Its sandy, loamy soil and moderate rainfall, and most of all its mountainous slopes that allow the water to run off easily and not soak the cinchona roots, bear a close resemblance to the hills on the shoulders of the trees' native Andes. Within a few years, *Cinchona ledgeriana* was growing all over the region. When the cinchona plantations in Java fell to the

Japanese during the Second World War, the only quinine that was available to the Allies came from the cinchona growers of the eastern Congo.

After the war, the Belgian plantation-owners would send their precious bark to the factory behind the white wall in Bukavu. There it was processed into totaquine, a powdery mixture of alkaloids with a quinine content of about 80 per cent that was packed in barrels and transported, first by truck via Bujumbura in Burundi to the west coast of Lake Tanganyika, then by ship across the lake, from there by train to Dar es Salaam, by ship to Hamburg and finally to Mannheim, where it was further refined into quinine sulphate. By the mid-1950s the heavy machinery for milling, cleaning and filtering the bark that had been bought second-hand from a British company was beginning to wear badly. Pharmakina asked Boehringer-Mannheim, the German manufacturer to which they sent most of the totaquine, to despatch a small team – two chemists, a laboratory technician and a metal-worker – to Congo to help restore its extraction plant.

Boehringer-Mannheim had a long association with quinine. As far back as 1837, a young German chemist named Conrad Zimmer began manufacturing quinine sulphate from quinine bark in Hesse. After the French chemists Pelletier and Caventou successfully isolated the quinine alkaloid from cinchona bark in 1820, enterprising chemists like Zimmer began manufacturing in many European cities. In due course, Zimmer's company was bought out by Engelmann & Boehringer, one of many small operations that that burgeoning family business took over in the course of the nineteenth century. In 1892 one of the Boehringer family's partners, Friedrich Engelhorn, became the sole owner of the company, though he kept the company name. Boehringer was rapidly becoming Germany's biggest manufacturer of quinine and quinine products, and under Engelhorn's leadership it had a new company seal and trademark designed, featuring a branch from the cinchona tree.

Shortly after its team refurbished the factory in Bukavu, Boehringer began making investments in Congo, eventually buying up the Pharmakina factory and, as more and more European plantation-owners left the country, the plantations themselves. Although chloroquine had taken over from quinine as the principal prophylactic against malaria, quinine was still being prescribed as a cure, and it was also widely used in the treatment of nocturnal cramp and the treatment of some heart diseases. By the mid-1970s, when the world price of quinine reached its peak, Pharmakina was processing more than three thousand tons of bark each year. Its Congolese plantations, not just of cinchona but also of coffee and tea, covered more than seven thousand hectares, and it employed close to ten thousand people. It liked to boast that its operation provided everything, from trees to treatment.

In the mid-1990s, though, three separate developments almost killed off the company altogether. Pharmakina's plantations were invaded by a fungus, a striped killer named *Phytophthora cinnamoni*, that insinuates itself into the space between the bark and the treewood, with the result that the tree weakens, dries out and eventually dies. First identified in Sumatra in 1922, the fungus has spread all over the world, but nowhere more than in Western Australia, where it affects nearly one in four tree species.

Phytophthora spread like bushfire through Pharmakina's Congolese plantations. The only economic solution was to pull up hectare after hectare of trees and leave the land fallow for three or four years while the fungus died out. But before the company had a chance to take any action, it was dealt another blow.

For decades, one of the principal markets for quinine had been as a cardiomuscular relaxant to help heart patients adapt to pacemakers. In 1995 a new product was launched on the market which, with none of quinine's side-effects, rendered the Jesuit powder irrelevant. Three-quarters of Pharmakina's market collapsed. Two years later, Boehringer-Mannheim was swallowed up by the Swiss pharmaceutical giant Roche.

The takeover meant a complete shift in strategy for the company. Instead of spreading itself around different areas of the industry, Roche decided to concentrate on becoming the world's leading developer of diagnostic treatments. What to do with Pharmakina was not initially a priority, but on 1 August 1998, when the rebel war began in eastern Congo between President Mobutu's soldiers and the Hutu *interahamwe* militias on one side, and the Tutsi-backed rebels who were fighting to overthrow Mobutu on the other, the Swiss mandarins decided to withdraw from the madness of trying to produce pharmaceuticals in Africa, and close down Pharmakina altogether.

Selling off the stock of totaquine held in Europe was the easy part; Amsterdam Chemie, the Dutch company that had traditionally refined all the *Cinchona ledgeriana* produced in Java, was happy to boost its own supplies. But dealing with the Congolese company, and the 1400 people it still employed, presented a greater problem.

For many weeks, Pharmakina's five European directors in Congo turned the matter over in their minds. Undertaking a management buyout of a sophisticated manufacturing operation in the midst of an African civil war sounded like a crazy idea. And there were other problems. It was one thing working for a German multinational, even at a distance; it was quite another owning and running a sizeable business as foreigners in a country that could be notoriously paranoid and nationalistic. The managers who took over Pharmakina would be risking not just their pensions, but perhaps their lives. And what about their families? What did their wives and children feel about committing themselves to a lifetime's investment in the heart of Africa?

In the end, two of the five top managers felt confident enough to take the plunge. Horst Gebbers, a German agronomist in his late fifties, had been in Congo for more than twenty years. His wife liked living there; moreover he had two sons who were eager to join the business, one in Germany and the other in Bukavu. Younger by more than a decade was a Belgian accountant, Etienne Emry, who had also lived in Congo for a long period; he was married, for the second time, to a Congolese wife.

In recognition of Boehringer's historic involvement with the quinine industry, Curt Engelhorn, Friedrich's grandson who was the head of Boehringer-Mannheim, committed family money to the management buyout. As important, if not more so in the years to come, was the moral support and the contacts within the pharmaceutical industry that the Engelhorns would offer the Pharmakina operation.

When Gebbers and Emry finally took over the business in the first few months of 1999, the war in Congo was going through a particularly nasty phase. The hundreds of thousands of refugees who had fled Rwanda at the time of the genocide were joined by thousands more whose lives had been overturned and their homes destroyed by the rebel forces intent on gaining power after the fall of Mobutu. Many of Pharmakina's plantations were cut off, and those that could be reached were not being properly tended. Production at the extraction plant in Bukavu slowed after the factory was attacked by rebels who caused millions of dollars' worth of damage to plant and machinery. Production was down to a tenth of what it had been five years earlier.

Pharmakina's new owners began a complicated restructuring. Crucial to their efforts was a redundancy plan that would reduce their workforce from 1400 full-time workers to fewer than five hundred. Each worker was offered a cash sum, but a far more important part of the package was the offer of a hectare of land, along with seed, fertiliser and logistical help. The offer of land close to where they lived meant that the workers would be able to feed their families and have some produce left over to sell in the market, and would not be displaced.

Once the corporate restructuring was complete, Pharmakina was able to concentrate on rebuilding its business. There was no fuel for sale locally, so to keep its trucks and factories running it had to import what it needed from Uganda, nearly a thousand miles away. Nor were there any banks, so Pharmakina's new owners turned to an Indian trader they had known for many years. Today, they pay him euros into a bank account in Europe, while he supplies them with local currency in cash. Which currency depends on which rebel group is in charge at any particular moment. At one stage, in the late 1990s, there were eight different currencies that could be preferred at any one time.

While the political situation remained unstable, Pharmakina concentrated on the plantations closest to the factory. It rebuilt roads and opened schools. Close by the main factory it built an AIDS clinic, the first of its kind in Bukavu. Without compulsory testing, Pharmakina has no idea what proportion of its workforce is HIV positive, but figures compiled by aid workers for the general population in the region show that it could be as high as one in four. By trying to treat, as well as educate, its workers, Pharmakina believes it is investing in the future.

In new plantations now being established within the virgin forest, Pharmakina clears the forest base for planting but does not cut down many of the bigger native trees. The system of husbanding natural resources in the field is as simple today as it was three centuries ago when the Jesuits were urging the Indians in the Andes to plant five trees for every tree cut down. Rather than in the shape of a cross, the cinchona trees in Congo are planted today in straight lines, two close together, then a third line of some taller species, such as *triphosia*, a leguminous shrub with fronds of leaves that provide the cinchonas with shade while they are young and still adapting to the environment.

The thick-stemmed cinchona trees that Ledger and Mamani saw in the Andean *cordillera* are rare today in Congo, although one that was planted by an Italian grower in Kaheyo in 1947 is still there to be gazed upon. For the most part, Pharmakina allows its trees to grow for seven or eight years before cutting them, that being the moment at which the quinine content is at its peak. The trunks are chopped down with *pangas*, heavy knives with wooden handles. They are then dragged out of the plantation to the roadside where they are stacked, waiting for a worker with a heavy mallet who will beat off the bark onto a dry cloth. Were Sir Clements Markham to find himself in Congo today, he would find nothing unfamiliar in the way the quinine bark is harvested.

After being dried in the hot sunshine for ten to fourteen days, the bark is taken to the factory in Bukavu where it is weighed, and its humidity and quinine content measured, before being shovelled into the hammer mill to be ground and then soaked with an acid base and slowly heated up. The crude quinine bisulphate that is produced is like hot coffee from which the grounds need to be strained. Once cool, the grounds provide much-needed compost for Bukavu's acid soil, while the liquid alkaloid is passed through a series of processes to clean, filter and crystallise it. In Boehringer-Mannheim's time the result was totaquine, the halfway-house product that would be shipped for refining in Germany. Under the new regime, the refining is done in Bukavu.

In a cupboard in Horst Gebbers' office are samples of the final products: quinine drops for babies, syrup for children (the taste of which is a distinct improvement on the Nivaquine syrup we were made to take on my grandparents' farm), and pills in elegant packs of blister foil. These come in two different chemical formulations and three different strengths. The only Pharmakina products that are manufactured abroad are the injectible ampoules, and that only as a precaution – injections made in eastern Congo, believed by many to be the AIDS centre of the world, were not thought to be easily marketable.

The success of Pharmakina's quinine operation lies, as it always has done, in the economics. The spread of malaria across this part of Central Africa is an example in miniature of the disease's expansion throughout the developing world. The deadly *falciparum* malaria that once was to be found in pockets across the continent now strikes along a broad belt that stretches as far south as Botswana and as far north as Chad. More than two hundred million people are at risk. The economic cost to these already impoverished nations is incalculable. And while resistance to chloroquine-based drugs has grown, the new drugs, such as Larium and Malarone, that have replaced chloroquine for Western travellers are so costly as to be unthinkable for Africans. A course of twelve Malarone tablets retails at more than $70. The Chinese artemisin-based drug Cotecxin retails for $6, but is often taken in conjunction with a course of antibiotic, which adds heavily to the cost. A small bottle of refined quinine salts, made in European and American factories, costs $12 in London, and more once it has been transported to Africa.

Dirk Gebbers, the young son of the agronomist who opted for the management buyout, and the man who will probably take over the business, describes the options that a patient with malaria faces in Africa. With an average income of $10 a month, an African family with a child who is sick with malaria can either give the child quinine, or let him or her die. Expensive Western drugs are simply not an option. A course of quinine hydrochloride, Pharmakina's most sophisticated product, sells in Bukavu for less than $2, and is, according to the local doctors, just as effective. Little wonder, then, that Pharmakina, which manufactures five hundred tonnes of quinine products a year, sells every pill it can produce.

But Gebbers knows that the future is not built just on mathematics. Promising to show me a surprise, he leads the way towards a low white building by the side of the main factory. There, in the company of his chief plantsman Elisé Mudwanga, who has worked for Pharmakina for nearly three decades, he shows off what he calls *la cuisine*, the laboratory where he has developed micro-cloning of the plantations' healthiest trees to produce a new generation of plants that are guaranteed to be free of the deadly *Phytophthora* fungus.

The technique is common in Europe and America, but has proved difficult to carry out in Africa, where electricity supplies are intermittent at best and trained laboratory technicians are hard to find. As a precaution, Pharmakina began its laboratory experiments in Switzerland. But transporting seedlings from Europe to the Congo proved too difficult – as with Sir Clements Markham's efforts more than a century earlier, most of the plants died en route. If Pharmakina was to have any success at cloning its trees, Gebbers realised, it would have to do it in the Congo.

In the white-walled *cuisine*, technicians separate a healthy cinchona leaf into pieces that are hardly a millimetre wide. Dropped into a cup of algal extract, with added sugar, vitamins, minerals and a plant growth hormone, the cinchona cells develop within three months into what is recognisably a tiny plant, with stem, leaves and pale translucent roots. Left for another three months in the rich algal extract, the cinchona seedlings are then ready to be planted out in the open, not in the forest yet, but in seedbeds that have been specially prepared to receive them. Regularly watered, the plants flourish in the warm African sun. After nine months they are ready for the open sea of the plantation.

Mudwanga reaches for a pot. In it is a seedling that is hardly bigger than my thumb. Its leaves show a busy network of narrow veins. Around its fragile stem, the bark covering where the quinine will develop is as fine as a silk stocking, but visibly already there. In that tiny plant, the eternal journey of the miraculous fever-tree has begun once more. In time the sap will swell, just as it did for the Jesuit priests who discovered it in another forest across another ocean. Sweet-tasting now, it will mature into a bitter brew as poisonous to the parasite as it is precious to mankind.

Introduction: The Tree of Fevers

Francesco Torti, who was such an enthusiast of the Peruvian bark that he turned his theory of treating illness into the engraving of the 'Tree of Fevers', published two treatises on fever, *Therapeutice specialis ad febres quasdam perniciosas*, in 1709, and a slightly enlarged version in 1712. Both manuscripts are in the Biblioteca Nazionale Centrale di Firenze. Torti's clinical consultations have been translated into English by that doyen of early cinchona studies, Saul Jarcho, and published in *Clinical Consultations and Letters by Ippolito Francesco Albertini, Francesco Torti and Other Physicians*, Francis A. Countway Library of Medicine, Boston, 1989.

Chapter 1: Sickness Prevails – Africa

Much of the material in this chapter is drawn from family letters, Gaston Bennet's inventory of my grandparents'1928 expedition to Africa, and the diaries that my grandfather, Mario Rocco, and my grandmother, Giselle Bunau-Varilla Rocco, kept for the years 1927–31 and 1939–45.

Edward Lear's letters were edited by Lady Strachey and published by T. Fischer Unwin in 1907. 'Malaria' appears in Giovanni Verga's first collection of tales about Sicily, *Novelle rusticane*, published in 1883 and translated into English by D.H.

Lawrence in 1925, three years after Verga's death. George Gissing, *By the Ionian Sea*, was published by Chapman & Hall in 1901.

Henry Morton Stanley's medicine chest, containing a single phial of the anti-malarial, 'Zambesi Rouser', was sold at Christie's, London, in November 2002.

Chapter 2: The Tree Required – Rome

Giacinto Gigli's diaries, 1608–70, are in the Biblioteca Vaticana in Rome. Alessandro Ademollo translated part of the manuscript into Italian in 1877, while Giuseppe Ricciotti undertook a fuller work in 1958. Neither is complete. Another contemporary chronicler of Counter-Reformation Rome is the Dutch diplomat Teodoro Ameyden (1586–1656), whose diaries were translated into Italian by A. Bastiaanse and published in 1967 as part of *Studien van het Nederlands Historisch Instituut te Rome* by Staatsdrukkerij, 's-Gravenhage.

Horace Walpole's letters are in the *Yale Edition of Horace Walpole's Correspondence*, 48 vols, 1937–82

For information about malaria in Italy, the fullest sources are F. Bonelli, 'La Malaria nella storia demografica ed economica d'Italia', *Studi Storici*, 1966, 659–721; Angelo Celli, 'The Campaign Against Malaria in Italy', *Journal of Tropical Medicine and Hygiene* 1908, 11,7: 101–8; and *The History of Malaria in the Roman Campagna from Ancient Times*, John Bale, Sons & Danielsson, 1933; Corrado Tommasi-Crudeli, 'On the Generation of Malaria in Flowerpots', *The Practitioner* 1881, 27: 3878 and *The Climate of Rome and the Roman Malaria* (translated by C.C. Dick), J. & A. Churchill, London, 1892; Ronald Ross, 'Dea Febris: A Study of Malaria in Ancient Italy', *Annals of Archaeology and Anthropology* 1909, 97–124. A more recent work is Robert

Notes on Sources

Sallares, *Malaria and Rome: A History of Malaria in Ancient Italy*, Oxford University Press, Oxford, 2002.

For detailed information on early medicine in Rome and the history of the Santo Spirito Hospital, see Louise S. Bross, *The Church of Santo Spirito in Sassia: A Study in the Development of Art, Architecture and Patronage in Counter-Reformation Rome*, Ph.D. dissertation, University of Chicago, 1994; Francesco Cancellieri, *Storia dei solenni possessi di sommi pontefici detti anticamente processi o processioni dopo la loro coronazione dalla Basilica Vaticana alla Lateranensa*, Rome, 1802; Alessandro Canezza and Mario Casalini, *Il Pio istituto di Santo Spirito e ospedali riuniti di Roma*, Rome, 1933; Pietro De Angelis, 'La Spezeria dell'arcispedale di Santo Spirito in Saxia e la lotta contro la malaria' in *Collana di studi storici sull'ospedale di Santo Spirito in Saxia, e sugli ospedali Romani*, # 13, Rome, 1954; Silvia De Renzi, 'A Fountain for the Thirsty and a Bank for the Pope: Charity, Conflicts and Medical Careers at the Hospital of Santo Spirito in Seventeenth-Century Rome', *Health Care and Poor Relief in Protestant Europe, 1500–1700*, edited by Ole Peter Grell, Andrew Cunningham and Jon Arrizabalaga, Routledge, London and New York, 1997; Camillo Fanucci, *Trattato di tutte le opere pie dell'alma città di Roma*, Rome, 1601; Eunice Dunster Howe, *The Hospital of Santo Spirito in Sassia and Pope Sixtus IV*, Ph.D. dissertation, Johns Hopkins University, 1978; Richard Krautheimer, *Rome: Profile of a City, 312–1308*, Princeton University Press, 2000; Laurie Nussdorfer *Civic Politics in the Rome of Urban VIII*, Princeton University Press, 1992; Angela Palma, *L'Ospedale di Santo Spirito in Saxia*, Biblioteca di Galileo, Padova, 1994.

The manuscript, now stolen, that Pietro De Angelis cites as containing the first mention of Peruvian bark is Fra Domenico Anda, Capo Speziale di Santo Spirito, *Ricettario manoscritto 334*, Biblioteca Lancisiana, Rome.

Chapter 3: The Tree Discovered – Peru

The most important sources of information that survive in Lima about Brother Agustino Salumbrino and the Jesuit College of San Pablo were gathered together by the Jesuit historian and collector, Father Rubén Vargas Ugarte. The two earliest volumes, *Libros de Viáticos y Almacén*, 1626–28 and 1628–31, are now in the Archivo General de la Nación (Companía de Jesús, Cuentas de Colegios, 1583–1669, Legajo 41, Cuaderno 2 & 3), as are the two later volumes, *Cuenta de la Botíca* 1757–67 and 1768 (Temporalidades Inventarios, Legajo 2, Cuaderno 28).

The correspondence between Brother Salumbrino and the Vicarate-General of the Order of the Society of Jesus can be found in the Archivio della Compagnia di Gesú in Rome, and details of some of the medicines he sent to Rome are available in the apothecary's archive which is to be found in the Archivio di Stato di Roma, La Sapienza.

On the early Jesuit presence in South America, Nicholas P. Custner, *Lords of the Land: Sugar, Wine and Jesuit Estates of Coastal Peru, 1600–1767*, State University of New York, Albany, 1980; S.J. Harris, *Jesuit Ideology and Jesuit Science: Religious Values and Scientific Activity in the Society of Jesus 1540–1773*, Ph.D. dissertation, University of Wisconsin, Madison, 1988; E. Saldamando Torres, *Los antiguos Jesuitas del Perú*, Imprenta Liberal, Lima, 1882; José Luis Valverde, *Presencia de la Companía de Jesús en el desarollo de la farmacia*, Universidad de Granada, 1978; Rubén Vargas Ugarte, *Historia de la Compania de Jesus en el Perú* (4 volumes), Burgos, 1963–65.

The biographical details about the Count of Chinchón come from his own memoir, *Memorias de los Virreyes del Peru*, edited and published in Lima in 1899, and from those of his secretary

J.A. Suardo, edited by Rubén Vargas Ugárte, *Diario de Lima (1629–34)*, Vasquez, Lima, 1935.

Early recountings of the legend of the Countess of Chinchón are to be found in Sebastiano Bado, *Anastasis cortices Peruviae, seu chinae chinae defensio*, Genoa, 1663; Gaspar Bravo de Sobremonte, *Disputatio apologetico pro dogmatica medicine praestantia*, 1639; Bernabé Cobó, *Historia del Nuevo Mundo*, Seville, 1890–95; Antonio de la Calancha, *Corónica moralizada del orden de San Augustín en el Perú*, Barcelona, 1638; Gaspar Caldera de la Heredia, *Tribunalis medici illustrations et observations practicae*, Antwerp, 1663; Pedro Miguel de la Heredia, *Operum medicinalium*, Lyons, 1665, Migual Salado Garcés, *Estaciones medicas*, Utrera, 1655.

For how the legend of the Countess should not be taken as fact, see A.W. Haggis, 'Fundamental Errors in the Early History of Cinchona', *Bulletin of the History of Medicine* 1941, 10,3: 417–59 and 568–92; Jaime Jaramillo-Arango, 'A Critical Review of the Basic Facts in the History of Cinchona', *Journal of the Linnaean Society of London (Botany)* 1949, 53: 272–311; C.H. La Wall, 'The History of Cinchona', *American Journal of Pharmacy* 1932, 104: 23–43.

For an analysis of the early writings on cinchona, see Francisco Guerra, 'The Introduction of Cinchona in the Treatment of Malaria', *Journal of Tropical Medicine and Hygiene* 1977, 80: 112–18 and 135–40; Fernando Crespo Ortiz, 'Fragoso, Monardes and pre-Cinchona Knowledge of Cinchona', *Archives of Natural History* 1995, 22,2: 169–81.

On early medicine in Peru, Juan B. Lastres, *Historia de la medicina peruana* (3 volumes), Universitad Nacional de San Marcos, Lima, 1951.

The most important work on the early trade across the Atlantic, Pierre and Huguette Chaunu, *Seville et l'Atlantique 1504–1650*,

Paris, 1955, shows that trade in cinchona did not exist in the early Conquistador period. For details of how the cinchona trade blossomed in the seventeenth and eighteenth centuries, see F. Arriba Arraz, *Papeles sobre la introducción y distribución de la quina en España*, Valladolid, 1937, which gathers together all the holdings regarding cinchona in the Spanish Archivo General de Simancas; also Edoardo Estrella, *Compendio histórico-medico-commercial de las quinas*, Burgos, 1992; John Fisher, *Commercial Relations Between Spain and Spanish America in the Era of Free Trade 1778–1796*, Liverpool, 1985, and Antonio Gonzalez Garcia-Baquero, *Cadiz y el Atlantico 1717–1778* (2 volumes), Escuela de Estudios Hispano-Americanos, Seville, 1976.

Chapter 4: The Quarrel – England

The account of the chartering of the *Conde* by, among others, Dona Maria Josepha de Albinar, Dona Pheliziana Barbara Garzan and Dona Elena Woodlock y Grant, and the cargo it transported from Guayaquil to Cadiz on its transatlantic voyage of 1748, is to be found in the Archivo General de Indias, Seville, Legajo 2770.

On Cardinal Juan de Lugo, see Camilo P. Abad, 'El Magisterio del Cardenal de Lugo en España con algunos datos mas salientes de su vida y siete cartas autografas ineditas', *Miscelenea Comillas* 1943, 1: 331–70; Gabriel Brinkman, *The Social Thought of John de Lugo*, Ph.D. dissertation, Catholic University of America, Washington D.C., 1957; J.S.I. Rompel, 'Kardinal de Lugo als Mazen der Chinarinde, Aus dem Leben das Kardinal', *75 Jahre Stella Matutina Festschrift* (volume I), Selbstverlag Stella Matutina, Feldkirch, 1931.

For details of the quarrel over the advisability of prescribing the Peruvian bark, see Joannes Jacobus Chifletius, *Pulvis febrifu-*

Notes on Sources

gus orbis americani ventilatus, Louvain, 1653; Antimus Conygius (pseudonym of Honoratus Fabri), *Pulvis peruvianus vindicatus de ventilatore eiusdemque suscepta defensio*, Rome, 1655; Vopiscus Fortunatus Plempius, *Antimus Conygius, Peruviani pulveris febrifugi defensor, repulses a melippo protimo belga*, 1665; Pietro Paolo Puccerini, *Schedula Romana*, Rome, 1651.

On how quinine influenced the shift in medical thought away from Galenic theory, see W.F. Bynum and V. Nutton (eds), *Theories of Fever from Antiquity to the Enlightenment, Medical History*, Wellcome Institute, London, 1981; Andreas-Holder Maehle, *Drugs on Trial: Experimental Pharmacology and Therapeutic Innovation in the Eighteenth Century*, Editions Rodopi, Amsterdam, 1999.

On malaria in England and the introduction of cinchona, A.W. Alcock, 'The Anopheles Mosquito in England', *Lancet* 4 July 1925, 208: 34–5; Leonard Bruce-Chwatt (with B.A. Southgate and C.C. Draper), 'Malaria in the United Kingdom', *British Medical Journal* 29 June 1974, 707–11; M.J. Dobson, 'Marsh Fever – The Geography of Malaria in England', *Journal of Historical Geography* 1980, 357–89, and 'Malaria in England: A Geographical and Historical Perspective', *Parasitologia* 1994, 36: 35–60; S.P. James, 'The Disappearance of Malaria from England', *Proceedings of the Royal Society of Medicine* 25 October 1929, 1–87; H. Kamen, 'The Destruction of the Silver Fleet at Vigo in 1702', *Bulletin of the Institute of Historical Research* 1966, 34: 165–73; Alice Nicholls, 'Fenland Ague in the Nineteenth Century', *Medical History* 2000, 44: 513–30; Paul Reiter, 'From Shakespeare to Defoe: Malaria in England in the Little Ice Age', *Emerging Infectious Diseases* 2000, 6.

The most important sources on Robert Talbor are his own writings, principally *Pyretologia: A Rational Account of the Cause and Cure of Agues*, London, 1672, and *The English Remedy: or*

Talbor's Wonderful Secret for Cureing of Agues and Feavers, Paris
and London, 1680 and 1682. See also George Dock, 'Robert
Talbor, Madame de Sévigné, and the Introduction of Cinchona:
An Episode Illustrating the Influence of Women in Medicine',
Annals of British Medical History 1922, 4: 241–7; S. Mason,
'Robert Talbor alias Tabor and the Ague', *Essex Journal* 1997, 32:
18–21; Rudolph E. Siegel and F.N.L. Poynter, 'Robert Talbor,
Charles II and Cinchona: A Contemporary Document', *Medical
History* 1962, 6: 82–5; Thomas Sydenham, *Observationes medicae*,
London, 1676.

On Charles II's health, Raymond Crawfurd, *The Last Days of
Charles II*, Clarendon Press, Oxford, 1909; Audrey B. Davies,
'The Virtues of the Cortex in 1680: A Letter from Charles Goodall
to Mr H', *Medical History* 1971, 15: 293–304.

On Oliver Cromwell's health, Leonard Bruce-Chwatt, 'Oliver
Cromwell's Medical History', *Transactions and Studies of the
College of Physicians of Philadelphia* 1982, 4,2: 98–121; Saul
Jarcho, 'A Note on the Autopsy of Oliver Cromwell', *Transactions
and Studies of the College of Physicians of Philadelphia* 1982, 4,3:
228–31.

Chapter 5: The Quest – South America

The original manuscript of John Evelyn's diaries, 1620–1706,
with the account of his visit to the Chelsea Physic Garden, is in the
British Library. E.S. de Beer edited them and a six-volume edition
was published by Clarendon Press, Oxford, in 1951.

Charles Marie de la Condamine's papers are in the
Bibliothèque Nationale in Paris. His *Journal du voyage fait par
ordre du roi* was published in Paris in 1751, and his account of the
discovery of the cinchona tree was published under the title 'Sur

l'arbre du quinquina', *Mémoires de l'Académie Royale des Sciences*, Paris, 1738. See also A. Le Sueur, *La Condamine d'après ses papiers inédits*, Paris 1911.

For an analysis of the growing importance of science to Spain, see Raquel Alvarez Peláez, 'La Historia natural en los siglos XVI y XVII', *La Ciença Española en Ultramar*, Madrid, n.d.; Barbara G. Beddall, 'Spanish Science and the New World', *Journal of the History of Biology* 1983, 16, 3, 433–40; Mauricio Nieto Olarte, *Remedies for the Empire: The Eighteenth Century Botanical Expeditions to the New World*, Ph.D. dissertation, University of London, 1994; M.L. de A. Turrion, 'Quina del nuevo mundo para la corona espanola', *Asclepio* 1989, 41: 305–24.

On Charles-Marie de la Condamine, Joseph de Jussieu and the first royal expedition to South America, see Antonio de Ulloa and Jorge Juan y Santacilla, *Relación histórica del viaje a le America meridional*, Madrid, 1748, and Joseph de Jussieu, *Description de l'arbre a quinquina, mémoire inédit*, Paris, 1936.

The principal sources of information on the second royal expedition to the New World are the *Journals* of Don Hipólito Ruiz, the manuscript of which was discovered by Jaime Jaramillo-Arango, then Colombian Ambassador to Great Britain, in the bombed-out basement of the British Museum during the Second World War. An English edition of the journals was published by Timber Press, Portland, Oregon, in 1998. See also Ruiz's *Quinologia o tratado del árbol de la quina ó cascarilla, con su descripción y las otras especies de quinos nuevamente descubiertas en el Perú*, Madrid, 1792, and its supplement.

On José Celestino Mutis and the Botanical Institute of New Granada, A.F. Gredilla y Gauna, *Biografía de José Celestino Mutis*, Madrid, 1911; E. Restrepo y Tirado, 'Apuntes sobre la quina', *Boletín de Historia y Antigüedades*, Bogota, 1943, 30; José Florentino Vezga, *La Expedición botánica*, Bogota, 1936; Alba

Moya Torres, *Auge y Crisis de la Cascarilla en la Audencia de Quito, Siglo XVIII*, Facultad Latinoamericana de Ciencas Sociales, Quito, 1994.

Chapter 6: To War and to Explore – from Holland to West Africa

Sources on the Walcheren expedition are surprisingly fragmented. The most informative are Brett James, 'The Walcheren Failure', *History Today* 1963/4, 13: 811–20 and 14: 60–8; Henry Coulburn, *Vicissitudes in the Life of a Scottish Soldier*, London, 1827; G.P. Dawson, *Observations on the Walcheren Diseases which Affected the British Soldiers*, Battely, Ipswich, 1810; William Dyott, *Dyott's Diaries, 1781–1845: A Selection from the Journal of William Dyott, sometime General in the British Army and aide-de-camp to His Majesty King George III* (edited by Reginald W. Jeffrey), Constable, London, 1907; Robert M. Feibel, 'What Happened at Walcheren: The Primary Medical Sources', *Bulletin of the History of Medicine* 1968, 42: 62–72; John Harris, *Recollections of Rifleman Harris as told to Henry Curling*, London, 1928; Hamilton, 'A Statistical Report of the Walcheren Fever, as it Appeared Among the Troops at Ipswich, on their Return from Holland', *Medical and Physical Journal* 1811, 25: 1–13; T.H. McGuffie, 'The Walcheren Expedition and the Walcheren Fever', *English Historical Review* 1947, 62: 191–202; Thomas Wright, President of the Medical Society and Temporary Physician to the Forces, *History of the Walcheren Remittant*, London, 1811.

On the exploration of Africa, Hugh l'Etang, 'Mungo Park', *The Practitioner* 1971, 207: 562–6; Michael Gelfand, *Rivers of Death in Africa: An Inaugural Lecture 10 October 1963*, Oxford University Press, Oxford, 1964; Robert M. Goldwyn, 'Medical Explorers in

Africa', *Journal of the American Medical Association* 7 April 1969, 208: 135–8; Marjorie Meehan, 'Physician Travelers', *Journal of the American Medical Association* 3 April 1972, 220: 97–102; W.E. Swinton, 'Physicians as Explorers: Mungo Park, the Doctor on the Niger', *Canadian Medical Association Journal* 17 September 1977, 117: 695–7, and 'Physicians as Explorers: David Livingstone, 30 Years of Service in Darkest Africa', *Canadian Medical Association Journal* 17 December 1977, 117: 1435–40.

The first serious compilation of statistics concerning the illness and death of soldiers serving in the British Army was *Statistical Reports on the Sickness, Mortality and Invaliding Among the Troops in Western Africa, St. Helena, the Cape of Good Hope and Mauritius*, 1840.

William Balfour Baikie's West African journey is recounted in *A Narrative of an Exploring Voyage up the Rivers Kwo'ra and Bi'nue (commonly known as the Niger and the Tsadda)*, London, 1856. His unpublished manuscript, *On the Remittant Fever of Western Africa*, 1856, is in the British Library, Add. Ms 32, 448.

Chapter 7: To Explore and to War – From America to Panama

On malaria in the United States, E.H. Ackernecht, 'Malaria in the Upper Mississippi Valley 1760–1900', *Supplement to the Bulletin of History of Medicine* 1945, 4; M.A. Barber, 'The History of Malaria in the United States', *Public Health Reports* 1929, 44: 2575–87; Elisha Bartlett, *The History, Diagnosis and Treatment of Fevers of the United States*, Philadelphia, 1847; A.W. Bennett, 'Malaria Endemic in Iowa', *Journal of the Iowa State Medical Society* 1943; J.R. St. Childs, *Malaria and Colonization in the Carolina Low Country 1526–1696*, Johns Hopkins University

Press, Baltimore, 1940; Daniel Drake, *Malaria in the Interior Valley of North America, 1850*, reissued by University of Illinois Press, Urbana, 1964; E.C. Faust, 'The History of Malaria in the United States', *American Scientist* 1951, 39: 121–30; L.W. Hackett, 'The Disappearance of Malaria in Europe and the United States', *Rivista di Parassitologia* 1952, 13:43–56; Margaret Humphreys, *Malaria: Poverty, Race and Public Health in the United States*, Johns Hopkins University Press, Baltimore and London, 2001.

On the use of quinine during the American Civil War, J.W. Churchman, 'The Use of Quinine During the Civil War', *Johns Hopkins Hospital Bulletin* 1906, 17: 175–81; and *Medical and Surgical History of the War of the Rebellion (1861–65)*, Part I by Joseph K. Barnes, 1870; Part II by Joseph Janvier Woodward, 1879; Part III by Charles Smart, 1888.

On the manufacture of quinine in America, J.W. England, 'The American Manufacture of Quinine Sulphate', *Philadelphia College of Pharmacy* 1898, 1: 57–64.

The most important sources on John Sappington are Sappington's own writings, *The Theory and Treatment of Fevers*, 1884, reprinted in 1971. See also Walter Morrow Burks, *Missouri Medicine Man*, M.A. thesis, University of Kansas City, 1958; Thomas Findley, 'Sappington's Anti-Fever Pills and the Westward Migration', *Transactions of the American Clinical and Climatological Association* 1967, 79: 34–44; Thomas B. Hall, 'John Sappington M.D. (1776–1856)', *Missouri Historical Review* 1930, 177–99; Lynn Morrow, *John Sappington: Southern Patriarch in the New West*, Columbia, Missouri, 1985; E.G. Riley, *John Sappington, Doctor and Philanthropist*, MA thesis, University of Columbia-Missouri, 1942; W.A. Strickland Jr., 'Quinine Pills: Manufactured on the Missouri Frontier', *Pharmacy in History* 1983, 25: 61–8.

For the medical history of the Panama Canal, see W.P. Chamberlain, *Twenty-Five Years of American Medical Activity on the Isthmus of Panama 1904–1929*, Panama Canal Press, 1929; J.F. Stevens, 'A Momentous Hour at Panama', *Science* 1930, 71: 550–2. The papers of Philippe Bunau-Varilla, George Goethals, William Gorgas and John Stevens are all in the Library of Congress, Washington D.C. The records of the Compagnie Universelle du Canal Interocéanique are in the National Archives of the United States, as are those of the Compagnie Nouvelle du Canal de Panama.

Chapter 8: The Seed – South America

The richest sources of information regarding the three most important plant hunters in South America are their own writings.

Charles Ledger's correspondence with John Eliot Howard is in the Royal Pharmaceutical Society of Great Britain along with the only known photograph of his Bolivian guide, Manuel Incra Mamani. See also Gabriele Gramiccia, *The Life of Charles Ledger (1818–1905): Alpacas and Quinine*, Macmillan, Basingstoke, 1986; John Eliot Howard, 'Origin of the Calisaya Ledgeriana of Commerce', *Pharmaceutical Journal* 1880, 10: 730–2.

Sir Clements Markham's private journals and his *Cinchona Notebooks (I and II)* are in the Royal Geographical Society, London. He also published three works on his travels in South America, *Travels in Peru and India while Superintending the Collection of Chinchona Plants and Seeds in South America and their Introduction into India*, John Murray, London, 1862; *A Memoir of the Lady Ana de Osorio, Countess of Chinchon*, Trübner, London, 1874; and *Peruvian Bark: A Popular Account of the Introduction of Chinchona Cultivation into British India 1860–1880*, John Murray, London, 1880.

Richard Spruce's letters and journals are at the Royal Botanic Gardens, Kew. His published writings include *Report (to the Secretary of State for India) on the Expedition to Procure Seeds and Plants of the Cinchona Succirubra, or Red Bark Tree*, 3 January 1862; and *Notes of a Botanist on the Amazon and the Andes* (edited by Alfred Russell Wallace), Macmillan, London, 1908, which begins with a long biographical section. See also M.R.D. Seaward and S.M.D. Fitzgerald (eds), *Richard Spruce (1817–1893): Botanist and Explorer*, Royal Botanic Gardens, Kew, 1986.

For the accounts of other travellers who went in search of the cinchona tree, see Alexander von Humboldt (translated by Aylmer Bourke Lambert), *An Account of the Cinchona Forests of South America Drawn up During Five Years Residence and Travels on the South American Continent*, Longman, London, 1821; Alexander von Humboldt and Aimé Bonpland, *Personal Narrative of Travels to the Equinoctial Regions of America During the Years 1799–1804*, Henry G. Bohm, London, 1852 and 1853; and Hugh Algernon Weddell, *Histoire naturelle des quinquinas*, Masson, Paris, 1849.

Chapter 9: The Science: India, England and Italy

On the chemistry of synthesising quinine, P. Pelletier and J. Caventou, 'Recherches chimiques sur les quinquinas', *Annales de Chimie et de Physique*, Paris, 1820, 15: 289–318 and 337–65.

On the hunt for explaining malarial infection, Amico Bignami, 'The inoculation theory of malarial infection', *Lancet* 3 December 1898, 1461–3 and 1541–4; Eli Chernin, 'Sir Patrick Manson: Physician to the Colonial Office 1897–1912', *Medical History* 1992, 36: 320–31; J. Franchini, 'Charles Alphonse Laveran', *Annals of the Association of Medical History* 1931, 3: 280–8;

Camillo Golgi, 'Sul ciclo evolutivo dei parassiti malarici nella febbre terzana', *Scienze mediche* 1889, xiii 7: 173–96; Charles Alphonse Laveran, 'The Pathology of Malaria', *Lancet* 1881, 2: 840–1; W.G. MacCallum, 'On the Haematozoan Infections of Birds', *Journal of Experimental Medicine* 1898, 3: 117–36; Dale C. Smith and Lorraine B. Sanford, 'Laveran's Germ: The Reception and Use of a Medical Discovery', *American Journal of Tropical Hygiene* 1985, 34: 2–20.

The correspondence between Sir Patrick Manson and Ronald Ross is in the library of the Wellcome Trust, London. A complete edition of the letters, edited by W.F. Bynum and Caroline Overy, was published under the title *The Beast in the Mosquito*, by Editions Rodopi, Amsterdam, in 1998. See also Patrick Manson, 'On the Nature and Significance of the Crescentic and Flagellated Bodies in Malarial Blood', *British Medical Journal* 8 December 1894, 1306–8; 'Surgeon Major Ronald Ross's Recent Investigations on the Mosquito-Malaria Theory', *British Medical Journal* 18 June 1898, 1575–7; 'The Mosquito and the Malaria Parasite', *British Medical Journal* 24 September 1898, 849–53; and Ronald Ross, 'On Some Peculiar Pigmented Cells Found in Two Mosquitoes Fed on Malarial Blood', *British Medical Journal* 18 December 1897, 1786–8; and 'Researches on Malaria: Nobel Medical Prize Lecture for the Year 1902', *Journal of the Royal Army Medical Corps*, April–June 1905, 88ff.

On the symptoms of malaria, S.T. Darling, 'The Spleen Index in Malaria', *Southern Medical Journal* 1924, 17: 590–5; H. Most, 'Falciparium Malaria: The Importance of Early Diagnosis and Adequate Treatment', *Journal of the American Medical Association* 1944, 124: 73–4.

Chapter 10: The Last Forest – Congo

Rubén Vargas Ugarte was very quick to mark the three hundredth anniversary of the discovery of quinine, with his article, 'Una fecha olvidada: El tercer centenario de descubrimento de la quina 1631–1931', which was published in *Revista historica*, Lima, in 1928, three years before the anniversary itself. For the festivities of 1931 see L. Suppan (ed.), *Celebration of the Three Hundredth Anniversary of the First Recognized Use of Cinchona*, Missouri Botanical Garden, St Louis, 1931.

On the establishment of the Dutch quinine plantations, see Norman Taylor, *Cinchona in Java: The Story of Quinine*, Greenberg, New York, 1945.

On malaria in the First World War, Sir S. Rickard Christophers, 'Malaria in War', *Transactions of the Royal Society of Tropical Medicine and Hygiene* 1939, 33: 277–304.

On malaria in the Second World War, Office of the Surgeon-General, US Army Medical Department, *Preventive Medicine in World War II: Volume 6, Communicable Diseases, Malaria*, Washington D.C., 1963. See also Mary Ellen Condon-Rall, 'Allied Cooperation in Malaria Prevention and Control: The World War II Southwest Pacific Experience', *Journal of the History of Medicine* 1991, 46: 493–513; A.F. Fischer, 'Cinchona in the Philippines', personal communication to Paul Russell, 1952; E. Haggerty, *Guerrilla Padre in Mindanao*, Longmans Green & Co., New York, 1946; Mark Harrison, 'Medicine and the Culture of Command: The Case of Malaria Control in the British Army During the Two World Wars', *Medical History* 1996, 40: 437–52; Robert J.T. Joy, 'Malaria in American Troops in the South and Southwest Pacific in World War II', *Medical History* 1999, 43: 192–207; J.S. Simmons 'The Army's Fight Against Malaria',

Journal of the American Medical Association 1942, 120: 30ff; W.C. Steere, 'The Botanical Work of the Cinchona Missions in South America', *Science* 1945, 101: 177–8.

For the account of the Japanese submarine, I-52, see the records of the USS *Bogue* and the account by Paul R. Tidwell, a maritime researcher, who found the wreck of the I-52 in 1994.

For the history of how quinine came to be grown in Congo, see Ernst Peter Fischer, *Selling Science: The History of Boehringer Mannheim*, 1992.

FURTHER READING

Cinchona and Quinine

Maria Luisa Duran-Reynals, *The Fever Bark Tree: The Pageant of Quinine*, Doubleday, New York, 1946

Anton Hogstad Jr et al, *Proceedings of the Celebration of the Three Hundredth Anniversary of the First Recognised Use of Cinchona*, Missouri Botanical Garden, St Louis, Missouri, 1931

Saul Jarcho, *Quinine's Predecessor: Francesco Torti and the Early History of Cinchona*, Johns Hopkins University Press, Baltimore and London, 1993

David Reynolds, *Steam and Quinine on Africa's Great Lakes*, Bygone Ships, Trains & Planes, Pretoria, 1997

Arthur Steele, *Flowers for the King: The Expedition of Ruiz and Pavon and the Flora of Peru*, Duke University Press, Durham, North Carolina, 1964

Disease

P. Ashburn, *The Ranks of Death: A Medical History of the Conquest of America*, Coward-McCann, New York, 1947

Philip D. Curtin, *Disease and Empire: The Health of European Troops in the Conquest of Africa*, Cambridge University Press, Cambridge, 1998

H. Cushing, *Life of Sir William Osler* (2 vols), Clarendon Press, Oxford, 1925

Gerald N. Grob, *The Deadly Truth: A History of Disease in America*, Harvard University Press, Cambridge, Massachusetts, 2002

J. Keevil, *Medicine and the Navy 1200-1900* (4 vols), Livingstone, Edinburgh and London, 1957–63

Paul de Kruif, *Microbe Hunters*, Harcourt Brace, Orlando, Florida, 1926

Mary Lindemann, *Medicine and Society in Early Modern Europe*, Cambridge University Press, Cambridge, 1999

Paul E. Steiner, *Disease in the Civil War*, Charles C. Thomas, Springfield, Illinois, 1968

Malaria in Literature

Wilkie Collins, *The Moonstone*, 1860

George Gissing, *By the Ionian Sea*, 1901

Henry Rider Haggard, *King Solomon's Mines*, 1886

Victor Hugo, *Les Misérables*, 1862

Henry James, *Daisy Miller*, 1879

Henry James, *The Portrait of a Lady*, 1881

Carlo Levi, *Cristo si è fermato a Eboli*, 1945

Jules Verne, *L'île mysterieuse*, 1870

Laura Ingalls Wilder, *The Little House on the Prairie*, 1935

Malaria in History

Leonard Bruce-Chwatt and J. de Zulueta, *The Rise and Fall of Malaria in Europe: A Historico-Epidemiological Study*, Oxford University Press, Oxford and New York, 1980

Robert S. Desowitz, *The Malaria Capers: Tales of Parasites and Peoples*, Norton, New York, 1993

Gordon Harrison, *Mosquitoes, Malaria and Man: A History of Hostilities Since 1880*, John Murray, London, 1978

Mark Honigsbaum, *The Fever Trail: The Hunt for a Cure for Malaria*, Macmillan, London, 2001

J. Jaramillo-Arango, *The Conquest of Malaria*, William
Heinemann, London, 1950

Roy Porter, *The Greatest Benefit to Mankind: A Medical History
of Humanity from Antiquity to the Present*, HarperCollins,
London, 1997

Ronald Ross, *The Prevention of Malaria*, John Murray,
London,1910

Ronald Ross, *Memoirs*, John Murray, London, 1923

Ronald Ross, *Memories of Sir Patrick Manson*, Harrison & Sons,
London, 1930

Paul F. Russell, *Man's Mastery of Malaria*, Oxford University
Press, Oxford and New York, 1955

James S. Simmons, *Malaria in Panama*, Johns Hopkins
University Press, Baltimore, 1939

Andrew Spielman and Michael d'Antonio, *Mosquito: The Story of
Mankind's Deadliest Foe*, Faber & Faber, London, 2001

Leon J. Warshaw, *Malaria: The Biography of a Killer*, Rinehart &
Co., New York, 1949

Christopher Willis, *Yellow Fever, Black Goddess: Plagues, their
Origins, History and Future*, HarperCollins, London, 1996

Exploration and War

Richard L. Blanco, *Wellington's Surgeon-General: Sir James
McGrigor*, Duke University Press, North Carolina, 1974

Philippe Bunau-Varilla, *Panama: The Creation, Destruction and
Resurrection*, Librairie Plon, Paris, 1903

Robert Bwire, *Bugs in Armor: A Tale of Malaria and Soldiering*,
Universe.com, Lincoln, Nebraska, 1999

Albert E. Cowdrey, *Fighting for Life: American Military Medicine
in World War II*, Free Press, New York, 1994

Richard B. Frank, *Guadalcanal: The Definitive Account of a Landmark Battle*, Random House, New York, 1990

M.D. Gorgas and B.J. Hendrick, *William Crawford Gorgas*, Lea & Febiger, Philadelphia, 1924

John Hemming, *The Conquest of the Incas*, Macmillan, London, 1970

David McCullough, *The Path Between the Seas: The Creation of the Panama Canal, 1870–1914*, Simon & Schuster, New York, 1977

A.B.C. Sibthorpe, *The History of Sierra Leone*, Frank Cass & Co., London, 1970 (first edition 1868)

Botany

Wilfrid Blunt, *In for a Penny: A Prospect of Kew Gardens*, Hamish Hamilton, London, 1978

G.H. Brockway, *Science and Colonial Expansion: The Role of the British Royal Botanical Gardens*, Academic Press, New York and London, 1979

E. St John Brooks, *Sir Hans Sloane: The Great Collector and his Circle*, The Batchworth Press, London, 1954

G. R. de Beer, *Sir Hans Sloane and the British Museum*, Oxford University Press, London, 1953

Richard Drayton, *Nature's Government: Science, Imperial Britain, and the 'Improvement' of the World*, Yale University Press, New Haven and London, 2000

Antonio Brack Egg, *Diccionario Enciclopedico de Plantas Utiles del Perú*, Centro de Estudios Regionales Andinos Bartolomé de las Casas, Cuzco, 1999

Victor von Hagen, *South America Called Them*, Robert Hale, London, 1949

Further Reading

Margaret Krieg, *Green Medicine: The Search for Plants that Heal*, Rand McNally, Chicago, 1964

Sue Minter, *The Apothecaries' Garden: A History of Chelsea Physic Garden*, Sutton, London, 2000

Toby and Will Musgrave, *An Empire of Plants: People and Plants that Changed the World*, Cassell, London, 2000

Toby Musgrave, Chris Gardner and Will Musgrave, *The Plant Hunters: Two Hundred Years of Adventure and Discovery Around the World*, Ward Lock, London, 1998

H. Waller (ed.), *The Last Journal of David Livingstone in Central Africa* (2 vols), London, 1874

INDEX

Index